SpringerBriefs in Climate Studies

SpringerBriefs in Climate Studies present concise summaries of cutting-edge research and practical applications. The series focuses on interdisciplinary aspects of Climate Science, including regional climate, climate monitoring and modeling, palaeoclimatology, as well as vulnerability, mitigation and adaptation to climate change. Featuring compact volumes of 50 to 125 pages (approx. 20,000- 70,000 words), the series covers a range of content from professional to academic such as: a timely reports of state-of-the art analytical techniques, literature reviews, in-depth case studies, bridges between new research results, snapshots of hot and/or emerging topics Author Benefits: SpringerBriefs in Climate Studies allow authors to present their ideas and readers to absorb them with minimal time investment. Books in this series will be published as part of Springer's eBook collection, with millions of users worldwide. In addition, Briefs will be available for individual print and electronic purchase. SpringerBriefs books are characterized by fast, global electronic dissemination and standard publishing contracts. Books in the program will benefit from easy-to-use manuscript preparation and formatting guidelines, and expedited production schedules. Both solicited and unsolicited manuscripts are considered for publication in this series. Projects will be submitted to editorial review by editorial advisory boards and/or publishing editors. For a proposal document please contact the Publisher.

More information about this series at http://www.springer.com/series/11581

Eric Ponthieu

The Climate Crisis, Democracy and Governance

Transition in Ten Steps: Action Points for Governments

Springer

Eric Ponthieu
Section for Agriculture, Rural Development
and the Environment—Sustainable
Development Observatory
European Economic and Social Committee
Brussels, Belgium

ISSN 2213-784X ISSN 2213-7858 (electronic)
SpringerBriefs in Climate Studies
ISBN 978-3-030-58129-9 ISBN 978-3-030-58127-5 (eBook)
https://doi.org/10.1007/978-3-030-58127-5

This Springer imprint is published by the registered company Springer Nature Switzerland AG
The registered company address is: Gewerbestrasse 11, 6330 Cham, Switzerland

Change will not come if we wait for some other person, or if we wait for some other time.
We are the ones we've been waiting for.
We are the change that we seek.

Barack Obama

Forewords

Climate change policies have been discussed globally for many decades now, most visibly starting in 1992. Without any doubt, much has changed since then: after the first careful steps in Kyoto (1997), a failed summit in Copenhagen (2009) then towards the groundbreaking summit in Paris (2015). But at the same time, our societies have, rightfully so, become impatient with policy-makers. Globally, the Fridays for Future strikes are calling us to finally convert the words into real climate action, something extra and effective that can achieve the targets that were established in the Paris Agreement.

As a Member of the European Parliament, I have been part of EU governance for more than 10 years. I have seen the debate changing, from fighting to be more ambitious in our 2020 targets to now, where the current European Commission has put the Green Deal centre stage for all of its policies. A huge shift in ambition, but again, words still need to be translated into real action. The very critical reactions of society on the Climate Law proposed by the European Commission show that the increased ambition is not yet fully converted into proposed action.

The gap between what science tells us, what society expects from us and what policy-makers are delivering is still huge. The Green Deal is an attempt to close that gap, but clearly, we are not there yet. While we understand the climate system better and better, it seems that less attention has been paid to understanding our governance system and why policy-making is lagging behind the needed and desired action. This book is aimed at filling in that gap: a necessary analysis of why policy-making is not delivering the actions that are so desperately needed and why it is so difficult to change the political system in order to mitigate the climate emergency.

For everyone interested in preventing further climate change, this book is a highly recommended to read, providing further insights into the complexities of our political system. It challenges complacency and provides a roadmap for how we can get change in our politics. This is the change we have been asking for during all those decades while climate science steadily progressed, showing that more delays in climate action will only make things worse. There is no planet B; let there be a political plan B.

Bas Eickhout, *Vice-President of the European Greens/EFA, Member of European Parliament since 2009, Brussels, Belgium.*

-o-

Climate change, climate crisis, existential crisis etc.: it sounds radical, but it is not. For decades, we have built a society addicted to fossil fuels, ruled by short-term policies and based on inequality.

These pillars that our system stands on are collapsing, and it is time to build new ones. Lots of world leaders have been trying to fix the system, instead of replacing the pillars. They have been fighting symptoms of a disease that they never really treated.

Although governments and politicians hold the prime responsibility to show ambition and bravery, they have been shamed by children who, in the last year, have sacrificed their education to protest against the world's inaction. Last September 2019, over 7.5 million people across the globe took to the streets, demanding to unite behind the science and provide a safe future.

Our rapidly disappearing carbon budget is our ultimate bottom line, revealed by the best available climate science. Scientists have written hundreds of reports, solutions and roadmaps to change the system and to fight for humanity and human equality.

In a world distracted by fake news, backroom politics and mainstream trivia, knowledge is key. While the world is facing huge challenges, many people are only aware of a little bit of what is happening. Even politicians are often lost in the dark. This is why more books like this one are absolutely necessary to educate people and equip them to fight for their right and what is needed. Also this book points to the responsibility of politicians and shows them there is a way ahead. No more empty words.

When people realise there is a way, they will fight for it. When they join us in this fight, things will change and frankly that is all everyone really needs. The world demands action. For things to change,

Anuna De Wever, *Young Climate Activist, Mortsel, Belgium.*

-o-

This is the book the next President of the USA needs to read, and also, the respective Presidents of the European Council and Commission, and also, political leaders of all countries where young climate activists have challenged their leaders to become climate leaders, And also, business executives who want to be able to look their children into their eyes in 2050, when the world must have reached carbon neutrality to respect the 1.5 °C warming limit. Last but not least, this is the book that must be read by climate-aware citizens who want to transform their anger into hope and then into action.

This is their handbook on how to address the climate change challenge. It is different from many 'how to save climate' books. Indeed, it focusses on the political leadership required to move the world economy in the direction needed.

Eric Ponthieu starting point is clear: 'Young people are perfectly right to shake existing governments'.

This book explains how public authorities can become climate leaders. How a new narrative that speaks to both the mind and the heart is needed. How we can use lessons from the handling of the COVID-19 crisis. How representative democracy can benefit from complementary deliberative approaches such as citizen's assemblies. How these approaches can help enshrine the long-term vision which is so often lacking in the political process. Eric Ponthieu pleads for improving the science–policy dialogue around tackling climate change, inspired by what happened during the COVID-19 crisis. He points to the importance of business having a long-term vision rooted in deliberations with its own employees and with civil society. He explains how sustainable consumption can be made mainstream and fashionable with the right government measures, reform of economic policies and support to local climate action aligned with national plans. Finally, he explains how all this could be funded, especially in Europe.

Eric Ponthieu warns the reader: civil disobedience due to mounting frustration in the face of climate inaction by governments is likely to increase. It has been non-violent up to now, but it is high time for real climate action to address youth and citizens' concerns. I agree.

Political leaders and business executives are human beings. They share our planet with almost 8 billion people. They often have children who will experience in their flesh the consequences of climate change and biodiversity loss in the coming decades. As the world economy will slowly restart after COVID-19, as huge amounts of money will be injected into the 'recovery', they have a unique chance to make a difference. They can recycle the old recipes, or reinvent a better world, cleaner and fairer, where the young people of today will be happy to live in. This book will inspire them.

Jean-Pascal van Ypersele, *Professor, UCLouvain, Former IPCC Vice-chair, Louvain-la-Neuve, Belgium.*

-o-

We are at a critical juncture for the climate crisis. COVID has exposed the fragility of many decision-making processes and much 'conventional wisdom' about the march of globalisation. And the recession is bringing the need for a 'fair' transition into sharp relief. COVID has also shown, however, how behaviours can change overnight. Amidst this disruption, the need to galvanise all sides to tackle climate change could not be greater.

This book provides a call to action and comprehensive set of steps for governments, businesses and communities to forge a new social contract. It explores how this can be shaped by ongoing deliberative engagement between people and decision-makers, including at a local level, in a way that looks for synergies and balanced outcomes. The implications of this go beyond climate to how to reconnect politics with long-term public interests.

Technological barriers to decarbonisation are small compared to the institutional, cultural and behavioural issues that need to be addressed. The book identifies the transformations in governance, partnerships and communications that can do this. It emphasises the importance of developing an inspirational narrative to help all sides understand that caring for the climate is good for the economy and society—as well as the natural world. Ensuring jobs and co-benefits resonate with local communities and making change easy and fun, so that people have a sense of agency and believe a positive future is possible, are essential for public acceptability.

Some companies are already considering what their 'Licence to Operate' means in a net zero world. However, there is more to do to ensure such changes shape all business models and core processes. The book poses important questions about the macroeconomic reforms that need to be discussed 'openly and without taboo' to create the frameworks for transformative change to be mainstreamed. How to measure well-being and internalise environmental costs into pricing—in a way that minimises the distributional impacts on those least able to pay—is key. A new set of principles for 'regulating for sustainability', including recognising the benefits of diversity and systems thinking, are needed.

The Youth Climate Movement has shown that creativity and different perspectives can unlock stale arguments and hold governments to account. With capacity building, the convening skills and soft power of NGOs can navigate complex, interconnected systems. The local networks, knowledge and partnership expertise of community groups are essential to enact the change we need to see.

COVID has taught us that the time for collaboration is now that we all have a role to play to ensure social—and natural—resilience. Climate change, like the pandemic, is far too important to leave to someone else to sort out. This book shows the way.

Sharon Darcy, *Director, Sustainability First, London, UK.*

Preface

Just as a generation of young people was taking to the streets to demand radical change on climate policies, the highways of Europe—and around the world—fell silent in response to the scourge of COVID-19. At every level, the extensive interventions being taken because of the life-threatening urgency of the pandemic contrast strongly with the uncertain and lethargic response of most governments towards climate change.

This book argues that at a time when the world is facing economic and social upheaval from a threat with the potential to affect everyone, there are vital lessons to be learned about our approach to the longer term but ultimately greater threat of climate change. In the latter case, the political establishment has been unwilling to listen and slow to act. Many of the actions taken to combat the spread and effects of coronavirus have required actions by governments and citizens unimaginable only a few months previously. Our democratic system itself has been willing to place safety and security ahead of some personal freedoms when it comes to public health. The response to COVID-19 has shown that if governments take a radical and progressive lead in making changes—imposing them if necessary—by creating a virtuous circle of positive and negative actions, there will be a hope of slowing, stopping and then reversing our global climate catastrophe. A ten-point manifesto explains how this can be done and how we can have hope for the future.

For the last 26 years, I have worked within the institutions of the European Union and seen concern about climate and the sustainability of our economic system steadily rise up the agendas of the European Parliament, Council and Commission. This concern has resulted in many attempts to equip the EU to face a different world—a world where the past arguments over, for example, energy, agriculture, transport or technology will be overtaken by one dominant question—by the end of this century will our planet be sustainable for humanity?

The response to the public health crisis of COVID-19 has reinforced my view that we do have, under certain circumstances, the technical, political and financial capacity to deal with pressing existential crises. I am much less confident about whether our collective understanding of climate change is adequate and therefore whether those steering our political processes will find the courage to take the necessary action to deal with an emergency wrongly perceived to have a much longer time span to deal with it. The global response to COVID-19 gives some grounds for both optimism and pessimism. However, it indicates that action can be taken urgently and at scale once risks are understood. Therefore, this book focuses on the necessary political leadership required, not least from the European Union itself, to follow up action on the pandemic with equally resolute action on the climate crisis.

Like the COVID-19 crisis, the climate crisis requires action by everyone. The book provides a realistic assessment of the support needed from society as a whole if leaders are to find the resolve to take action. Governments must lead and coordinate action domestically and globally but need the support of their citizens and the cooperation of business. We need massive social and economic change requiring collaborative effort, highlighting the benefits and sharing the burdens, because we face serious dangers. Our cohesion as a society is at risk as generations are differentially affected. COVID-19 might create a situation where climate issues could be shelved for some years as the world's economies try to return to business as usual. This cannot be allowed to happen.

This book is written for:

- **Decision-makers and influencers** in or connected with European governments, legislatures and political parties, at all institutional levels,
- **Institutional, corporate and non-governmental stakeholders** who are or should be in the process of defining their response to the climate crisis,
- **Individuals**, whether as consumers or climate warriors, young or old, nationalist or globalist,
- **Media and communication channels** of all types which will play a vital part in conveying the urgency and necessity of climate action.

This book proposes a ten-point manifesto of actions which will change the state of affairs and bring new life to climate governance and democracy. Starting with a long-term and integrated vision, it sets out steps to change our political and economic thinking through leadership, communication, partnership, dedicated institutions, local action, new economic models, lean consumption and cultural change. While there has never been such a pressing necessity for a 'third way', the book is an urgent call to governments to reform their way of deciding and implementing climate policies. It complements many essays on the technical measures and policies needed to reduce emissions. It addresses those frequently neglected political processes that lead to our democracies favouring the election of politicians prepared to take radical economic, social and political action on behalf of the planet. Its main message to everyone is, 'Don't despair, there is hope!' The success of the transition to a sustainable world depends on governments leading with a programme of long-term, comprehensive, integrated policies which are enthusiastically supported by the civil society as a whole.

Brussels, Belgium Eric Ponthieu

Acknowledgements

This book is the result of many years of conversations, meetings and reading. Many of my colleagues and friends have unconsciously fed the internal reflection process that led me to start gathering facts and figures and sketch the early ideas. But what drove me to organise, shape and express this wealth of material was the admirable way young people from all continents defended the common cause of climate protection and the lack of effective response from politicians to this call. At all stages of the book's development, I constantly returned to those faces of young people in the street to keep me motivated in realising this project. I want to pay homage to the dignity with which they are defending a cause that ought to be the cause of everyone and that ought to become the central goal of policy-makers in the next decades.

My closest inspirers are from the academic world, universities and research centres alike. The list of those who gave me the initial impetus or forced me to reflect differently on an issue that is now mainstream in the media is long. However, I particularly want to warmly thank Dr. Sander Chan, Senior Researcher at the German Development Institute, Prof. Seiji Ikkatai from Musashino University, Prof. Izumi Inasawa from Ritsumeikan University and Eamon O'Hara, Executive Director of ECOLISE.

To this incomplete list, I want to add one person who has been the main provider of inspiration and motivation, Richard Adams, whose career has spanned many activities which have tried to make the world a better place and who was a former member of the European Economic and Social Committee. Without Richard's help, this book would not have become real. His experience in publishing and in developing systemic thinking on emerging issues has been instrumental to my completion of the writing. With his current responsibilities in the fair trade movement and the care of older people he brought very beneficial and complementary perspectives to my own thinking. The outcome of this last year's companionship with Richard will not be limited to one supplementary book on the shelves of central libraries but also to the start of a long-lasting friendship.

Last but not least, I want to thank those who have spent considerable time in reading and commenting on intermediary drafts, and in particular, my friends Georgios Amanatidis, Sébastien Occhipenti, Ovidiu Romosan and Frederic Sgarbi. And my last words are for my family, my spouse, Danielle, and my children, Coline, Valentin and Arthur, who all supported me in their own way to finalise a project they felt was very important to me.

Eric Ponthieu

Supportive Statements

"This book is for everyone who is searching for a visionary and yet pragmatic guide to overcoming the economic, political and cultural barriers to action on the climate crisis. Acknowledging the various interests at stake, Dr. Eric Ponthieu maps out how to get all the different stakeholders to unite for the shared purpose of striving for a climate-neutral society. The reader should feel encouraged to draw from the elaborated holistic and cross-sectoral measures for transformation, to help co-create the envisioned future that honours climate justice and seeks to enable the well-being and inclusion of all."

—Amelie Krug, *Youth Climate Activist, Student of Cultural Anthropology and Social Sciences at Humboldt University of Berlin.*

"Displaying the inability of the current political system to tackle the upcoming challenges, Dr. Eric Ponthieu's book highlights the critical necessity of a collective action for transforming our current society, it's economic and political heritage. While calling for a strong governmental leadership, the book draws on the inspiration from community-led actions and astutely recognises the importance of empowering local communities."

—Alisa Sidorenko, *council member of ECOLISE (European metanetwork of community-led initiatives on climate change and sustainability), board member of Suderbyn Permaculture Ecovillage, Sweden.*

"Dr. Eric Ponthieu's call for a democratic transformation comes particularly timely. As governments deal with multiple crises, including a global health emergency, accelerating and dangerous climate change and an economic recession, new and radical governance approaches need to break with incrementalism and respond to the calls for action by a new generation. By highlighting action by governments, Eric Ponthieu refocuses on the critical role of public authorities in driving democratic change."

—Dr. Sander Chan, *Senior Researcher at the German Development Institute/ Deutsches Institut für Entwicklungspolitik (DIE); Adjunct Assistant Professor at Copernicus Institute of Sustainable Development, Utrecht University.*

"This book, in focusing on the important role of public authorities in dealing with climate change and a move to a low carbon economy, is a call to action at a critical juncture. Local councils can play a key role in consensus building in their respective communities. In my own town awareness and action has come about as a result of our sustainability action plan, declaring a climate emergency, promoting improved biodiversity and encouraging the community as a whole to adopt a more sustainable intergenerational approach."

—Robert Hull, *Mayor of Hexham, Northumberland, UK, former Director, European Economic and Social Committee, Head, Environment Policy Division, European Commission.*

Structure of This Book

This book is an address to policy-makers and decision-makers in the form of a manifesto. Made up of the ten most necessary changes in the way to decide and apply climate action (defined as climate governance), it starts first with a summary of the ten key action points which is then followed by chapters which set the contemporary scene relating to the climate crisis, the demand for radical changes coming from young people and civil society and the EU developing policy plans. Then the ten action points are fully expanded with reference to this context. This layout enables the readers to either decide to immediately access the full description of key points at the end or go through the entire contextual details for fully grasping the relevance of the ten action points.

Contents

Chapter 1
A Transition Manifesto: 10 Action Points for Governments

The world as we have created it is a process of our thinking. It cannot be changed without changing our thinking.
Albert Einstein

Abstract Not only is climate change a fast-growing issue that affects the ecosystem and its capacity to deliver services, it is starting to weigh heavily on the very foundations of society. Although the general public remain largely under- and ill-informed scientific evidence is accumulating and the more aware in civil society increasingly feel worried about the impacts of climate change. All governments, whatever their past credentials on green issues, are facing increasing pressure to place climate action at the centre of their programme. Yet there has been very little practical action, a stark contrast with the speed and substance of decisions taken to respond to the COVID-19 pandemic.

The world is in transition—but to what? If, at the end of this century, humanity is in a secure position, with hope for the future, it will look back at the period between 2020 and 2050 and see that this was a time when the world changed course to deal with existential problems. This book focuses on the climate threat but the massive global response to a life-threatening virus already has shown the strengths and weaknesses of government action and public response. It is likely that those future generations will call these three decades the Transition, a term that has become common currency amongst those recognising the need for change. This can only be done with the active engagement of citizens who understand that a new social and economic contract is required. Effective political leadership is vital. Governments—the governments at different institutional levels and of Europe in particular—are therefore requested to take full account of the following ten proposals which are aimed at substantially enhancing climate governance in their developing climate and economic plans and making the Transition a reality.

E. Ponthieu, *The Climate Crisis, Democracy and Governance*,
SpringerBriefs in Climate Studies,
https://doi.org/10.1007/978-3-030-58127-5_1

Action Point Summaries

Ambitious States in the Driving Seat

Protecting Mother Earth, just as protecting citizens from the threat of a novel, highly contagious virus, requires governments to assume strong and ambitious leadership. They will need to place climate mitigation and adaptation at the centre of their policy and their action. Governments' leadership must take different forms. Listening to scientists and the concern from the streets, they must unroll the necessary measures for an environmentally sustainable transformation of society that generates positive responses from its social and economic constituents.

Instead of modestly attenuating the negative impacts of market driven economies, they must seize the initiative and establish a new economic system in the form of a well-being economy that brings benefits to all. At the heart of the required political decisions is the obligation to change lifestyles in a way that is relatively painless and positively socially endorsed. Decisions on setting limits must be fair and balanced, and governments cannot shy away from taking regulatory measures to limit the most unsustainable practices. They must incentivise climate action domestically, with the challenging task of engaging all sections of society and must also do so internationally by stimulating the take up of equally ambitious measures elsewhere.

A climate leader will be someone who can embrace all these challenging tasks at once and turn them into positive and coherent political action.

Turning Climate Protection into a Consensual Transformation of Society

Changing mind sets and turning apathy into enthusiasm to act for the climate needs a new narrative that speaks to both the mind and the heart. Unlike the immediate threat from COVID-19 the climate crisis, though as real and ultimately more devastating, seems less urgent. Taking inspiration from the European Commission proposal for a Climate Pact, a massive, decade-long communication campaign has to be launched without delay. The European Green Deal (EGD) summarises its proposed benefits as cleaner air, water, soil and energy with more charging points for e-cars; reusable or recyclable packaging and less waste; better public transport alternatives; renovated homes, schools and hospitals; less pesticides and fertilisers; healthier food and more environmentally-friendly products in our shops; and better health for current and future generations. All these positive features need to be re-packaged into targeted messages and their urgency and necessity emphasised by reference to the failure to prevent the pandemic taking hold and its subsequent devastation, even though such an event was recognised as probable.

Climate action ambassadors, selected from civil society for their reaching-out potential, will have to explain, promote and enrich the simple and more substantial daily gestures that reduce carbon footprints. Social media should be the main vector of diffusion of this Transition narrative. This communication campaign must be a cross-party endeavour that is robust enough to deal with legitimate political challenge as well as the propaganda of the climate change deniers. The core messages of the campaign should inspire all to act positively for the climate. Turning fear or passivity into enthusiasm should not be the exclusive responsibility of the ambassadors and politicians. All business, organisations and indeed every stakeholder in society should be encouraged to play their part.

Partnering with Civil Society and Citizens to Co-elaborate and Co-implement Policies

All organisations of civil society (public and private organisations, non-governmental organisations, religious organisations etc.) must be engaged in the collective endeavour to salvage and restore Mother Earth. The ability to mobilise civil society has already been clearly demonstrated in response to the pandemic. Citizens will do more themselves if their voice really counts in supplementary democratic processes between elections. Representative democracy must be complemented with deliberative approaches such as citizens' assemblies in order to enlarge the scope of popular perspectives, solutions and to co-design policies. Those governments introducing new life to the democratic processes will reap almost immediate results: more endorsed policies, more effective reductions of carbon emissions and a reduced societal cost for the region or the country. More public deliberations are also a means to prevent social polarisation which could limit the capacity of society to make a unified stand for climate protection.

Developing Capacity to Deliver the Long-Term Vision

Precedence must be given to a long-term vision for a sustainable, pro-climate transformation of society. This vision should prevail over the sectoral policies that may emerge during the journey. Even if pressed by climate urgency, good climate governance means giving enough time to formulate and elaborate efficient policies that recognise and resolve potential problems such as stranded assets, redundant workers and neglected regions. The long-term vision should be backed by a wide political consensus and even be enshrined into the primary law of countries. Citizens must be encouraged, inspired and educated to give this long-term vision the required importance when voting at elections. There will be natural links to be made with the numerous supportive and remedial actions being taken in the postcorona world, and these can be used to stimulate engagement by civil society.

Developing Integrated and Holistic Approaches Underpinning the Long-Term Vision

Besides developing a long-term vision, governments have to develop the capacity to optimise the climate response by activating a large range of sectoral policies. The horizon of possible policy combinations and synergies can be expanded by creating dialogues with citizens and representative organisations from the different sectors involved. A closer science-policy relationship is needed to address the additional complexity derived from elaborating integrated policy approaches. The scienceled strategies used to tackle the COVID-19 outbreak have strengthened the place and importance of evidence-based action. In accordance with action point 2, holistic policy-making has to be properly communicated to all concerned to assure that the complexity of the political proposal does not blur the clarity of the joint endeavour.

Mobilising the Business World

The corporate sector has a key role to play in partnership with government and civil society and its response to the numerous challenges of the pandemic showed that such a focus on the public good is possible. The time has come to work hand in hand with the business sector to accelerate the climate efforts, moving away from incremental change to transformational change. The most forward-looking companies already know they do not have a secure future beyond 2050. Private organisations have to apply to themselves the good governance principles of developing long term, integrated pro-climate visions based on civil society deliberations and engagement. This means going beyond an ambitious programme of corporate social responsibility (CSR) and Responsible Business Conduct (RBC), by implementing a fundamental change in business models and processes, notably via a special focus on Environmental, Social and Governance (ESG) factors. In this way the private sector could be one of the main winners from the substantial changes in the offer and demand for climate friendly products and services. The main drivers here are the sustainable economic models such as the circular economy, the sharing economy and the transformation of a product-based economy to a service-based economy. People, not only as citizens but specifically as employees, should be closely involved with the preparation and implementation of new climate-friendly business approaches. Governments have to incentivise, through regulation, norms and sector-specific accompanying measures, the internalisation of climate protection within corporate affairs. It is also necessary that the extensive aid given to the corporate sector in response to COVID-19 is linked to a new global social contract based on clear, measurable, corporate standards of conduct.

Financing the Transition

As the pandemic has shown, an existential crisis requires a lot of money to deal with it and the good news is that this money exists. The creation of a European climate bank may enable the financing of the transformation required by the Paris Agreement. It has to be supported by EU governments and its creation has to be fast-tracked. The private sector, financial investors, fundraising companies and charities have to be incentivised through regulation to redirect their funds to climate protection. The new EU taxonomy to qualify "green funds" and the recourse to environmental, social and governance (ESG) factors should reduce greenwashing attempts and give an extra boost to sustainable investment funds that are already steeply rising. Fund raising and "climaction" campaigns should also serve to widely communicate the multiple gains that a society can reap from caring for future generations.

Acting Effectively at the Local Level

The vital nature of local networks and community response will already have proved themselves in the solidarity and effectiveness of the response to the covid public health threat. Marshalling the whole of society behind the collective endeavour of positive climate action can be best realised by engaging stakeholders at the local level, where the reality of climate change impacts and the benefits of action can be directly observed. To ensure effective action at the local level, climate governance must be both multi-level and multi-stakeholder. The different governmental levels (central, regional, local) have to create propitious conditions for local climate action to be aligned with national plans. Governments and regions also have to support local and sub-local multi-stakeholder networks bringing together communities, citizens and civil society. They have to encourage the vanguard regions, cities and local communities who go faster than central governments towards climate neutrality.

Paving the Way for a Reform of Economic Policies

Today's economic system based on GDP (Gross Domestic Product) has to be replaced by a new environmental and social macroeconomics. COVID-19 may possibly have rewritten the rules but it has certainly made us ask what rules are the most important. We have the opportunity to radically re-think the unsustainable and unequal global growth model and replace it with an emphasis on well-being, sustainability and equity. By providing a much wider picture of the state of development of countries, the new macroeconomics will enable governments to take full account of the climate threat in the design of the entire policy spectrum. To orient policies towards more inclusive and qualitative prosperity, governments must dare to regulate the excesses of the market, both supply and demand. Major corporations of all types have an understandable tendency to equate their own interests with that of the common good when this is not necessarily so. The business of the near future can only be legitimized to function by the state when its objectives align with the values and requirements of the society in which it operates. Economic policy reform must de-legitimise excessive and unequal wealth accumulation through a wide range of taxation measures, notably income tax progressivity Greater social and economic equity is the foundation for greater collective climate action. At global level, particularly in the context of international trade, travel and tourism, products and services must have their full external environmental cost included, with additional and progressive disincentives on carbon hungry activities.

Making Sustainable Consumption Mainstream and Popular

Helped by the large communication campaign under action 2 of the manifesto, governments have to review their entire regulatory arsenal to give preference to the consumption of products and services that are produced in a sustainable manner. To prevent unsustainable consumption from continuing to proliferate, governments have to set clear limits that dissuade this form of unethical and irresponsible consumption and this will go hand-in-hand with the new attitudes needed in deal with the outcome of the global pandemic. Governments have to move away from assuming that efficiency gains deriving from product innovation will lower the overall energy consumption without looking at the changes in the way people consume these products. The rebound effect must be addressed through renewed behavioural, legal and economic instruments, and in particular through a fair energy taxation that does not penalise low income households.

Endangered Climate—Endangered Humanity

Climate change must be seen as a truly global issue that weighs heavily on every aspect of social life and whose pressure will not be reduced by current approaches. The "act now or pay more later" refrain has been heard for years but either we have not believed the science or we have decided to pass the bill to future generations …

Every day, there is someone, somewhere, saying that climate change is going to be the biggest challenge ever faced by humanity. Even when faced with other existential challenges, the Executive Secretary of the United Nations Framework Convention on Climate Change (UNFCCC) Patricia Espinosa warned when announcing the postponement of the November 2020 climate change summit (COP26)[1] in Glasgow:

[1] Conference of the Parties.

"COVID-19 is the most pressing threat to humanity today, but we must not forget that climate change is the greatest threat in the long term". Scientists, and in particular the International Panel on Climate Change (IPCC), have been saying it ever more loudly since 1988, the date of the creation of their grouping under the auspices of the World Meteorological Organisation (WMO) and of the United Nations Environmental Programme (UNEP).[2] The First World Climate Conference was held even earlier, in 1979, in Geneva with scientists from 50 nations. Even then it was agreed that alarming climate change trends made it necessary to act urgently. Similar and more worrying concerns have increasingly been outlined by the IPCC and other groupings of scientists until the present, especially at the 1992 Rio Summit, and through the 1997 Kyoto Protocol the 2015 Paris Agreement. Even if the calls for action by scientists[3] are not always unanimously shared, they all stress the same messages: ambitious action is needed now, delay at your peril.

Although short-term reductions in pollution and greenhouse gas (GHG) emissions resulted from the COVID-19 pandemic, only long-term systemic shifts will change the trajectory of carbon dioxide (CO2) levels in the atmosphere. The trend is that GHG emissions are still rapidly rising, as evidenced by the last UNEP Emission Gap report[4] which shows that, in the past decade they have risen by 1.5% a year on average. The UNEP warns that "unless global GHG emissions fall by 7.6% each year between 2020 and 2030, the world will miss the opportunity to get on track towards the 1.5°C temperature goal of the Paris Agreement". And, says the UN agency, among the G20 nations, which account for 78% of all emissions, only 5 countries (including the EU) have committed to a timeline for meeting net-zero emissions.

Current endeavours have to be increased, by an order of magnitude, to conserve the habitability of our biosphere, as highlighted in the special 2018 IPCC report on the impacts of global warming of 1.5 °C above pre-industrial levels.[5] Evidence is accumulating that the consequences of climate change are already affecting and will increasingly affect every nation and all peoples on our planet. No country, no person will escape from its dramatic impact which will take different forms in different parts of the globe. The last five years are on track to be the warmest on record. Alarmingly, global temperatures are set to increase by at least 3 °C towards the end of the century—twice what climate experts have warned is the limit to avoid the most severe economic, social and environmental consequences. Polar ice caps are melting and the sea is rising. Glaciers are shrinking whilst ice on rivers and lakes is breaking up earlier. Plant and animal ranges have shifted and trees are flowering sooner and are becoming more vulnerable to new kinds of disease. Biodiversity as a whole is in decline with the ongoing sixth great extinction of planetary life. Extreme weather events, alternating devastating droughts and increased rainfall are becoming more common. The destruction of the structural pillars of our ecosystem

---

[2]History of the IPCC https://www.ipcc.ch/about/history/. Accessed 28 May 2020.

[3]Ripple et al. (2019).

[4]UNEP, Emission Gap Report 2019, November 2019, p. 108.

[5]IPCC report, Global warming of 1.5 °C, October 2018.

(Ice caps, rainforests, tundra and marine systems) is accelerating to a point of no-return. Furthermore, our social cohesion, which is already under threat, is facing new challenges. It is increasingly likely that in order to adapt to the consequences of climate change, more and more people will attempt to migrate to safer and greener pastures in the coming decades. The list of potential suffering is painfully long.

In addition, scientists point to two categorical messages—delay and impact. Already in its fourth Assessment Report dating back to 2007, the IPCC said very clearly that the costs of inaction are much greater than the costs of action. In other words any delay will make matters worse and harder to resolve. The ticking of the clock becomes louder and louder. Postponing only means paying much more financially and socially at a later stage. If a decrease in carbon emissions had begun with COP15 in 2009,[6] we would not have to reduce them by an extra 7.6% from 2019.[7] And if nothing substantial is done by 2022, we will have to reduce carbon emissions by 10% every year thereafter.

The second scientific warning concerns the intensity of the negative consequences of climate change. New studies and investigations made by scientists always present a more gloomy picture of climate change's impacts compared to previous studies.[8] The December 2019 report of the European Environment Agency (EEA) on the 'State of the Environment'[9] states that Europe faces environmental challenges of unprecedented scale and urgency. Even though the report recognised that Europe has made a positive start in tackling its own carbon emission, water pollution or plastic waste, it will not achieve its sustainability vision of 'living well within the limits of the planet' by continuing to promote economic growth and seeking to manage the derived environmental and social impacts. A study published in May 2020 in the Proceedings of the National Academy of Sciences of the United States of America (PNAS) review[10] says that GHG emissions will threaten the life of a third of the world population in the next 50 years. Around the Mediterranean, the average temperature, typically 13 °C in 2020, could be as high as 20 °C in 2070.

In its 2020 Global Risks Report, the World Economic Forum (WEF), that is made up of the world's top corporate leaders, recognises that climate change is striking harder and more rapidly than many expected. For the first time in the history of the Global Risks Perception Survey, environmental concerns dominate the top long-term risks by likelihood among members of the WEF's multistakeholder community. Three of the top five risks by impact are also environmental. "Failure of climate change mitigation and adaption" is the number one risk by impact and number two

[6]COP15 in Copenhagen was initially presented as the COP where bold decisions would be taken. This only happened 6 years later at COP21 in Paris.

[7]UNEP, Emissions Gap Report 2019, November 2019.

[8]New climate models predict a warming surge: www.sciencemag.org/news/2019/04/new-climate-models-predict-warming-surge. Accessed 28 May 2020.

[9]EEA, The European environment—state and outlook 2020: knowledge for transition to a sustainable Europe, 2019.

[10]Xu et al. (2020).

by likelihood over the next 10 years, according to our survey. The Forum's multi-stakeholder network rate "biodiversity loss" as the second most impactful and third most likely risk for the next decade.

The bill for climate change rises daily. For some, the phrase 'climate change' no longer seems appropriate to reflect the urgency of the challenge. Since May 2019, the British daily newspaper, The Guardian, has been using 'climate emergency' instead, arguing that climate change is too neutral. This reflects the view that climate change is the symptom of a more generalised malfunctioning of our ecosystem. Some scientists[11] believe that the core environmental problem is the excessive use of natural resources—the overshoot. The non-governmental agencies (NGO) Global Footprint Network (GFN)[12] and World Wide Fund for Nature (WWF) measure the overshoot via the environmental footprint of countries and calculate their overshoot day: the date when a country, through production and consumption, has fully depleted its own annual quota of environmental resources. Any overshoot day before the 31st of December means that the country lives on credit at the expense of the planet by trimming the capital of natural resources. Though their method of calculation cannot be considered as 100% accurate in absolute terms, the way the overshoot day has evolved in the last 20 years reveals a lot about our rapacious way of grabbing finite environmental resources. The earth overshoot day has moved up the calendar by two months over the past 20 years. In 2019, it was on 29 July, the earliest date ever. To reach the end of 2019 it was necessary to consume 1.75 earths to satisfy humanity's needs in food, energy, housing, etc. Developed countries are consuming considerably more resources than developing countries. The EU had its 2019 overshoot day on 10 May, 2.5 months before the world average.[13] According to GFN, not a single EU country is performing at a sustainable level. Moreover, developed countries absorb large amounts of resources coming from other, less developed, countries.

Another sign of the global character of the environmental crisis has been provided in 2000 by Nobel prize winner Paul Crutzen, with co-author Eugene F. Stoermer,[14] who have coined a new name for describing the current geological epoch, the Anthropocene. They emphasised the noticeable global effects of human activities since the first industrial revolution, confirming the central role of mankind in geology and ecology.[15] They proposed the latter part of the eighteenth century for the start of Anthropocene, as it is the period when data retrieved from glacial ice cores show the beginning of a growth in the atmospheric concentrations of several greenhouse gases, in particular CO_2 and methane (CH_4).

If climate change has historically been investigated in close relation to other environmental issues, a more recent trend consists of bringing into the loop social, economic and even political issues. The word Capitalocene has for example been

---

[11]Heinberg (2017).

[12]https://www.footprintnetwork.org/. Accessed 28 May 2020.

[13]Global Footprint Network and World Wide Fund for Nature, ed. World Wide Fund for Nature, April 2019EU overshoot day—living beyond nature's limits.

[14]International Geosphere-Biosphere Programme (IGBP), newsletter 41.

[15]Steffen et al. (2011).

used as an alternative to Anthropocene to refers to the ravages of capitalism, not only on climate change but on the whole of society.[16] According to Jason W. Moore, the global crises of the twenty-first century are rooted in the age of capital, and not just in the age of man. Another, intensely political viewpoint has come from the French researcher, Malcolm Ferdinand, who prefers to speak of Negrocene to connect climate change with the (post) colonial question. Arguing that the exploitation of nature has gone hand in hand with the exploitation of black people, he calls for a decolonial ecology.[17]

These different interpretations are painting the problem of climate change on a wider and more diverse canvass. They have the merit of informing governments of the absolute need to address climate change from a comprehensive, inter-policy perspective. Even if the link between climate change and the social security system may seem tenuous at first, there is value in understanding how the two interact. It is the same for health issues and COVID-19 has brought home to us in a dramatic and catastrophic way the reality that the management of nature is closely linked with the management of human health.

The lesson for governments is as clear as the warnings from science: forget about locating climate change as a subset of the matters dealt with by an Environment Minister—it must be a part of everyone's portfolio.

Climate Protection in the Shadow of the COVID-19 Pandemic

The response to the COVID-19 pandemic has shown the importance of a cross-cutting perspective and co-ordinated global action

The global health crisis engendered by the spread of the novel coronavirus, COVID-19 has had an immediate, worldwide impact in almost every sphere of life. Until March 2020, climate change was increasingly presented in the media as the most worrying issue faced by humankind. At this point, it was replaced by the urgent reality of a global pandemic. There was no debate about the need to respond urgently and effectively in the face of the direct and immediate threat to human life—action followed quickly. Within a few weeks, far-reaching decisions were taken and over 75% of the world's population was under some form of a lockdown. Normal life was suspended and this response rapidly created a dilemma. Would the devastating economic consequences of putting modern economies into a coma create a vacuum of climate ambition and action in the future? The debate continues and remains unresolved but there are grounds for hope.

During the covid crisis, the climate has not been ignored. On the one hand, a great reduction in industrial and urban pollution became noticeable across the world when countries went into lockdown. At the same time, research showed that

[16] Moore (2016).

[17] Ferdinand (2019).

of the coronavirus deaths across 66 administrative regions in Italy, Spain, France and Germany, 78% of them occurred in just five regions, and these were the most polluted.[18] Although there is no absolute certainty about the link between environmental degradation and zoonotic diseases, many studies identify possible environmental links with the origin, spread and impact of the virus. According to the executive director of UNEP, Inger Andersen, nature is sending us a message with the coronavirus pandemic and the ongoing climate crisis.[19] Exploitation of wildlife by humans through hunting, trade, habitat degradation and urbanisation facilitates close contact between wildlife and humans, which increases the risk of virus spillover.[20]

A very unexpected and welcome impact of the pandemic has been to refresh the political debate on changing the DNA of our economic model. The covid crisis, with its diversity of metaphors (war, plague, tsunami, etc.) has crystallised a number of the problems of our failing economic and social system and of the negative effects of globalisation (excessive profit-seeking, runaway de-localisation, over-exploitation of natural and human resources, etc.). The collateral victims of the pandemic were so many and its destructive reach so extensive that massive state intervention and support was the global response. This was a crisis that silenced even the most ardent defenders of the power of the market, not only in the health sector but in the economy as a whole.

Questioning dogma brings hope but the challenge will be to build convincing arguments that sustainable alternatives cannot be limited to an opposition between liberalism and nationalism. The reconstruction of a "new world" where some previous certainties have been overturned will require the creation of a new economic 'software'. Succumbing to the temptation of relaunching the economy without modifying the main operating system risks leading to a multiplication of systemic crises in the medium term. In the same way that the post-war global economic framework was predicated on the creation of supranational institutions such as the UN, the International Monetary Fund and the World Bank, there is a need to reform existing institutions and establish new ones, notably for enabling the EGD to meet its climate neutral objective in the most socially and economically efficient way.

There are many voices arguing that the pandemic has created the opportunity to set the world on a new, sustainable course. This was notably demonstrated with EU environment ministers arguing that the EGD "must be central to a resilient recovery after COVID-19". This was also firmly stated by top U.S. and British economists, including Nobel laureate Joseph Stiglitz from Columbia University and prominent British climate expert Lord Nicholas Stern.[21] More than 700 economic stimulus policies launched during or since the 2008 financial crisis were examined by more than 200 experts from 53 countries. There is plain evidence that green

---

[18] https://www.theguardian.com/environment/2020/apr/20/air-pollution-may-be-key-contributor-to-covid-19-deaths-study. Accessed 29 May 2020.

[19] https://www.theguardian.com/world/2020/mar/25/coronavirus-nature-is-sending-us-a-message-says-un-environment-chief. Accessed 28 May 2020.

[20] Johnson et al. (2020).

[21] Hepburn et al. (2020).

projects such as boosting renewable energy, energy efficiency in the building sector, R&D, workers reconversion or vocational training, create more jobs, deliver higher short-term returns and lead to increased long-term cost savings relative to traditional stimulus measures. This group of economists support massive programmes of green public investment as the most cost-effective way both to revive virus-hit economies and strike a decisive blow against climate change. The call is an invitation to governments to move quickly from emergency economic relief measures to green recovery measures.

Comparing the threats of COVID-19 and climate change is revealing. Both are seen as global, existential threats but there are notable differences and also similarities. Unlike COVID-19 significant action to analyse climate change had been underway for decades and global action plans for dealing with it had been debated at the highest international level and some steps taken to implement them, notably the Paris Agreement. But like COVID-19 the 'solution' to climate change was seen by many as necessitating significant economic adjustment. Two things are clear. Firstly, the fundamental causes of climate change will remain long after the world has adjusted to the coronavirus pandemic. Secondly, in addition to globalisation increasingly dysfunctional ecological and environmental trends enabled the novel coronavirus to arise and rapidly spread.

In many respects, the coronavirus pandemic is a re-run of the 1918–20 influenza virus pandemic. There is strong evidence that both resulted from animal to human viral transmission.[22] In 1918, the influenza virus was thought to have originated in the farm state of Kansas, "where dirt-poor farm families struggled to do daily chores—slopping pigs, feeding cattle, horses, and chickens, living in primitive, cramped, uninsulated quarters".[23] It's not known whether it started in the pigs or chickens or birds flying overhead. But it spread to young farmers who, drafted for World War I, reported for duty at Fort Riley.'[24] From there it went to Europe and thence, in several waves, globally, hastened by the dispersal of armed forces at the end of the First World War.

The COVID-19 crisis could be regarded as a devastating warning and wake-up call to heed the larger crisis of the climate threat. Even as the pandemic spreads, it is probable that a combination of social mitigation measures and medical research will bring the virus under control. We will have made major short-term changes, previously unimaginable, to the way we live and we will then be facing a significant but not insuperable political and economic challenge. The climate crisis is of a different nature. It is potentially far more devastating and its 'solution' is even more complex, involving action across almost every sphere of life. But the response to the pandemic will have changed our mindset. We will be more aware of risk in our interconnected world, have a greater understanding of the necessity for global cooperation, have discovered that, in times of crisis, unimaginable sums of money can be found, and that co-ordinated, evidence-based government action on scientific

[22]https://www.ncbi.nlm.nih.gov/pmc/articles/PMC6218637/. Accessed 29 May 2020.

[23]https://www.kansas.com/news/local/article200880539.html. Accessed 29 May 2020.

[24]https://www.ncbi.nlm.nih.gov/pmc/articles/PMC340389/. Accessed 29 May 2020.

advice can solve urgent problems. A new awareness of the importance and role of nature and, hopefully, a new attitude of caring for each other and of the value of life could make smoother the pathway to dealing with the challenges of climate change.

Resulting from the pandemic, the importance of early, science-led, co-ordinated global action has become apparent. The need for decisive government action underpinned by the informed consent of the public is also an important lesson. The radical, short-term changes we have made in response may encourage us to make the long-term changes needed to deal with the climate emergency.

Are People Fully Aware of the Implications of Climate Change?

By and large, the general public is still under-informed about climate change and its consequences. This hampers any far-reaching and collective action which would stave off the climate crisis. Additionally, misinformation abounds – not just from the threatened minds of climate-deniers but more subtly from those with an interest in preserving the status quo.

Year by year, the number of people having a clearer vision of what climate change is about and what it implies in terms of impacts and adaptations grows, but there is still much to be done to ensure that a minimal understanding of the consequences of climate change is widespread. In a 30-country survey mandated by the French utility EDF[25] and undertaken by the polling company IPSOS in early 2019, 23% of the countries' respondents doubt that climate change's cause is anthropogenic whereas 8% do not believe that climate change even exists (up to 19% in the USA). Not only are nearly a third of people unconvinced that humans are responsible for runaway climate change there remains much confusion between climate change on the one hand, and air pollution and depletion of the ozone layer on the other hand.

Scepticism touches all social classes, and especially those who feel that their vested interests could be seriously affected by the required transformation. The ignorance and demagogy of some politicians has fueled contempt for education and science. Naomi Oreskes, an American historian of science, says in her book *Why trust science?*[26] that rejection of science is not correlated with education but with ideology, drawing similarities between the climate change debate and earlier controversies over smoking, acid rain and the hole in the ozone layer. This highlights the need to both improve the understanding of people about climate change and offering them ways to develop a personal stake in a decarbonised society.

---

[25]Climate and Public Opinions International Observatory, Guillaume Petit, EDF and IPSOS, November 2019, https://www.edf.fr/en/opinions-observatory-global-warming. Accessed 29 May 2020.

[26]Oreskes (2019).

Returning to the IPSOS survey it was also the case that 74% of respondents were more worried about climate change than they were 5 years ago, showing the importance of recent media influence. In 2019, there was hardly a day without the climate crisis hitting the headlines. Concern is rising, particularly amongst the young. In Europe (43 vs. 30%[27]) and in North-America (44 vs. 27%) of young people are more worried than the rest of the population; they have more to lose and longer to live with the consequences of that loss.

Regarding the adequacy of the current remedies a large majority of people think that the signatory countries will fail to respect their commitments under the Paris Agreement. For 70% of respondents, governments should be the first to act, far ahead of citizens (45%) and businesses (32%). The reason seems to be that people understand that regulation is essential when, in a crisis, citizens fail to act. This does not prevent 53% of respondents thinking that climate protection comes about principally through a change of lifestyle—and 55% have already taken steps in doing some action themselves. While vested interests tend to push responsibility to the governments, there is however a strong and legitimate citizens' demand for instruction and enforcement to come from "above" given the very wide scope of required individual change.

The persisting inertia of governments in handling the climate issue has triggered a recent demand for a change in terminology. For Jean-Yves Pranchère, Professor of political philosophy, the terms 'climate warming' or 'climate change' have not been sufficient to stimulate consciences because of their soothing character and their failure to reflect the disruptive effects that are involved.[28] The call to use the term "climate emergency" has come not only from the school climate strikers but also from different civil movements such as Extinction Rebellion that was founded in the UK in October 2018. In the scientific world, there is no unanimity on changing the description. What is plainly agreed is that even with a temperature increase of only 1.5 °C, there will be certain parts of the world which will be very severely impacted, where 'emergency' will become disaster. Speaking of a climate emergency or climate crisis is relevant for those regions that are already being violently affected and, in time, the rest of the world will follow.

In part, the appeal for a change of terminology has been heard. On 1 May 2019, British Members of Parliament endorsed a Labour motion to declare a formal climate and environment emergency. Ireland, Canada, France and Spain followed. In November 2019, the European Parliament (EP) voted by a large majority in favour of a resolution declaring climate emergency in Europe. The timing was meant to pile pressure on the EU's new European Commission to deliver an ambitious European Green Deal, which it did in December 2019. At the local level, the movement has even been more responsive with tens of large and small cities (New-York, Paris, Mulhouse, Lyon, Napoli, Milano, Heidelberg, Bonn, Karlsruhe, Bern, Geneva etc.)

---

[27]The second figure is for those over the age of 55.

[28]Le Soir, Brussels, 27 June 2019, p. 18.

adopting a climate emergency. By April 2020 1482 jurisdictions in 28 countries with a total population of 820 million had declared a climate emergency.[29]

It is too early to say whether and how the change of wording will affect the overall message about the climate and its perception by the public. Scaring people by more dramatic and realistic wording can have unwanted consequences although it seems to have been effective in the case of the covid crisis. Nevertheless, the risk exists that the new description of the climate issue plunges people into a sort of paralysis with climate protection being equated with survival. According to Michael Mann, director of the Earth System Science Center at Pennsylvania State University, "this doom and gloom attitude leads people down a path of despair and hopelessness and finally inaction, which actually leads us to the same place as outright climate-change denialism." Mann also argues that we cannot assume that climate change deniers continue to promote false information and use deflection tactics such as encouraging people to make minor lifestyle changes while neglecting the need to press for serious policy reform.[30]

With the multiplication of fears, there is also the return of war as a background preoccupation. This was particularly obvious in the 'battle' against COVID-19. The German philosopher Hans Jonas thought that fear can be part of the solution, by making people more rational.[31] Fear is not only a negative emotion, he said, as it may be assumed as a positive virtue. However, could such a positive reaction be expected to happen in the current environmental crisis which is predominantly marked by the diffusion of more and more frightening messages and which has such a global scope that no individual, group or country can have any meaningful impact? In social media, there are more and more people exchanging messages of despair and resignation, creating a generalised feeling of insecurity and distrust in their institutions In these circumstances it is not guaranteed that a grassroots movement for change will arise spontaneously. Nevertheless, the immediate reality of the COVID-19 crisis, affecting everyone's lives and where fear was undoubtedly a stimulus to action, has been a reminder that events forecast by experts as highly probable and yet dismissed as unlikely by governments and the general public can, indeed, happen. The months in which the covid crisis developed could be seen as an accelerated prefiguration of the two or three decades in which the chaos contingent on climate change would develop. When faced with these kinds of risks, individual and collective reactions can be rejection and apathy but also a determination to act collectively as greatly demonstrated during the covid crisis.

To motivate the general public to act, a more subtle, and positive type of message will have to be crafted. What to tell people and how to tell it are questions requiring urgent consideration if the general public is to be mobilised in a supportive way - as proposed in the first action point of this manifesto.

---

[29] https://climateemergencydeclaration.org/. Accessed 29 May 2020.

[30] https://www.theguardian.com/science/2019/nov/09/doomism-new-tactic-fossil-fuel-lobby. Accessed 29 May 2020.

[31] Jonas (1980).

The pandemic is likely to have made people more receptive to the warnings of scientists and to the need for prompt action.

References

Ripple WR, Wolf C, Newsome TM, Barnard P, Moomaw WR, World Scientists' Warning of a Climate Emergency, BioScience, 5 November 2019.

Chi Xu, Kohler T, Len T, Svenning J, and Scheffer M, Future of the human climate niche, PNAS, https://doi.org/10.1073/pnas.1910114117. Accessed 28 May 2020.

Heinberg R. (2017) There's No App For That. Post Carbon Institute, Santa Rosa.

Steffen W; Grinevald J; Crutzen P; McNeill J, The Anthropocene: conceptual and historical perspectives, Philosophical Transactions of the Royal Society A: Mathematical, Physical and Engineering Sciences, **369** (1938): 2011, 842–867.

Moore, JW (2016) Anthropocene or Capitalocene?: Nature, History, and the Crisis of Capitalism, PM Press, Oakland.

Ferdinand M (2019) Une écologie décoloniale, Le Seuil, Paris.

Johnson C K et al. Global shifts in mammalian population trends reveal key predictors of virus spillover risk, Proceedings of the Royal Society B, 8 April 2020, https://doi.org/10.1098/rspb.2019.2736. Accessed 28 May 2020.

Hepburn C, O'Callaghan B, Stern N, Stiglitz J, Zenghelis D, Will COVID-19 fiscal recovery packages accelerate or retard progress on climate change?, Oxford Review of Economic Policy, 8 May 2020.

Jonas H (1980) ed. Kranzberg M, The Heuristics of Fear, in Ethics in an Age of Pervasive Technology, Westview Press, Boulder.

Oreskes N (2019) Why trust science? Princeton University Press, New Jersey.

Chapter 2
The European Green Deal and Other Climate Plans

Abstract Politicians promote bright ideas, using terms referring to glorious past initiatives, but implementation is another matter. With its European Green Deal, the EU has mapped out an interesting pathway but its climate neutrality goal still seems out of reach if it is to be achieved through the proposed outline measures. EU decision makers know there is a high risk in addressing lifestyles and hyper-consumerism and remain unwilling to do so. The lack of political leadership is pushing an ever larger portion of European society, and not just climate activists, to ask for changes now. A possible way to avert the growing civil disobedience movement is to get both politicians and people mobilised by bringing issues together in win-win policy packages.

Many climate plans are now on the drawing board, with some partially recognising the scale of the challenge by referencing major social planning initiatives of the past century. The proposal for a European Green Deal is the most far reaching attempt to put climate protection at the core of EU policy-making. Will commensurate and consistent implementation follow suit?

Among the burgeoning visions for new or upgraded climate plans, the "New Deal" terminology predominates—first coined by Franklin D. Roosevelt for his programmes to combat the Great Depression in the USA. The words Green New Deal are used by Jeremy Rifkin,[1] the American social thinker, who defends a pro-market and pro-technology approach. His belief is that the reduction in the cost of renewable energy will come to our rescue. By 2019 solar generation technology had become cost-competitive with that of oil, coal and nuclear even though it faced specific constraints of deployment. Referring to the advent of the third industrial revolution (coal and oil as being instrumental in the first two), the result would be a huge market shift around 2028 that would sweep fossil fuels away, once and for all. For this to happen, huge investments in digital, smart and decentralised infrastructure

[1] Rifkin (2019).

E. Ponthieu, *The Climate Crisis, Democracy and Governance*,
SpringerBriefs in Climate Studies,
https://doi.org/10.1007/978-3-030-58127-5_2

would obviously be needed. But the price that has to be paid is worth it as, on top of creating a massive shift to a post carbon ecological era, Jeremy Rifkin also predicts the onset of a "social capitalism" where the market would play the role of a "guardian angel of humankind".[2]

The vision articulated by the US Representative Alexandria Ocasio-Cortez, a Democrat candidate for the 2020 presidential election in the USA, together with Senator Ed Markey, has similarities to that of Rifkin but is undeniably of a wider scope. The starting point is the same: huge investments in renewable energy, a definitive stopping of fossil fuel subsidies and a carbon neutrality by 2050. But on top of these measures there are also key proposals to revitalise other important sectors of activity such as education and health, with finance coming from the wealthiest sector of society. The US Representative sees her plan, above all, as a great opportunity to address climate change while also correcting economic inequalities. Interestingly, the public support for the plan significantly increased during the covid crisis, moving from 47% in May 2019 to 59% in April 2020.[3]

Taking the same line as Alexandria Ocasio-Cortez is Naomi Klein. In her latest book, "*On Fire*,"[4] she defends a vision of a Green New Deal that builds upon the hope instilled by the climate strikes of young people. She does not believe in the omnipotence of markets. Instead, she positions individuals as the main drivers of the Transition. Climate change is more than a crisis of excess emissions, she says, it is also a crisis of mismanagement of natural and human capital resources. At stake is the preservation of our democratic systems. The economy is wider than mere energy change and she feels that we have been waiting for a market-based energy transition for too long. Also, she does not believe in a technological revolution as some of the required natural resources will not reduce, quite the contrary if one takes the resources needed in producing batteries. Today our dependence on rare earths and rare metals is huge, which has led the French journalist Guillaume Pitron to refer to China, the major world producer of these resources, as "a Saudi-Arabia of rare metals".[5]

Naomi Klein thinks the environmental revolution must also imply a willingness to turn our back on the fundamentalism of the free market economy.[6] The government's role in her Green New Deal is immense as they would have to undertake severe restrictive measures such as limiting extraction or preventing resource overuse and wastage by the richer segments of society. She denies that it may lead to an austere

---

[2]https://www.climatechangenews.com/2019/06/03/finland-carbon-neutral-2035-one-fastest-targets-ever-set/. Accessed 31 May 2020.

[3]https://www.euractiv.com/section/climate-environment/interview/oslo-mayor-heres-how-we-plan-to-become-a-carbon-neutral-city/. Accessed 31 May 2020.

[4]Klein (2019).

[5]Pitron (2018).

[6]Klein (2019).

future as the restricting measures will be counterbalanced by more justice, more collective deliberations and more contentment. She sees her proposal as a valid project as "it does not ask people to choose between the end of the month or the end of the world".

For the American philosopher Michael Walzer, the Green New Deal proposed by his US compatriots should actually refer to a Green Social Democracy, because of the ecology and society nexus. For him, the new social democracy must be green and also respond to the aspirations and worries of the workers.[7]

A similar reflection has happened on the other side of the Atlantic with the EGD proposed at the beginning of her mandate by President of the EC, Ursula von der Leyen, in December 2019. Acknowledging that she had been greatly inspired by the passion of millions of young people demonstrating in the streets, she identified the EGD as the first flagship proposal of the new EC's mandate. The EGD is a transversal policy proposal aimed at putting the EU on track to become the first carbon neutral region of the world from 2050. The major innovation is that, if finally adopted by the EU co-legislators (European Council and European Parliament), this central objective will reconfigure the entire legislative and regulatory arsenal of the EU. It is made up of ten specific measures that cover almost all EU sectoral policies: industry, finance, taxation, R&D and innovation, energy, transport, infrastructure, agriculture, food etc. Proposals for each of these policy areas will be brought forward throughout 2020–21.

As a first key measure, the EC adopted in March 2020, the first climate legislation to enshrine the 2050 climate-neutrality target in law. The EU policy-making efforts to reduce CO2 emissions started in 2009 with the so-called 20/20/20 package: 20% reduction of CO2 emissions by 2020 compared to 1990 levels, 20% of renewable energy used as a total of energy consumption by 2020 and a 20% increase in energy efficiency by 2020. The EU is on the right path to achieve the target of reducing GHG emissions by 20% below 1990 levels by 2020, as evidenced by the decoupling of economic activity from GHG emissions: between 1990 and 2017, the combined GDP of the EU increased by 58% while total GHG emissions fell by 22%.[8] In October 2014, the EU decided to further reduce its GHG emissions by 40% in 2030.[9] The EGD proposal implies the upgrading of the current goal of 40% reduction by 2030 which will be the subject of a further assessment in the course of 2020. The figure of a minimum of 50% seems probable, the EP, having called in its October 2018 resolution for an increase of the EU GHG emission reduction target of 55%.

The EC has committed to review every EU law and regulation to align them with the climate neutrality objective. The extension of the EU Emissions Trading System (ETS) to cover the maritime sector and reduce the free allowances allocated to airlines over time, as well as the introduction of a carbon border adjustment

---

[7]Interview in the Belgian newspaper Le Soir, 29 October 2019, p. 18.

[8]Amanatidis (2019).

[9]A policy framework for climate and energy in the period from 2020 to 2030, COM/2014/015 final. https://eur-lex.europa.eu/legalcontent/EN/TXT/?uri=CELEX:52014DC0015. Accessed 31 May 2020.

mechanism to discourage imports that would not respect the Paris Agreement and prevent carbon leakage, are among the proposed measures. Other foreseen measures concern the revision of the energy taxation directive, the further greening of the common agricultural and fishery policies, and the integration of the EGD into free trade agreements concluded by the EU. However, as every bit of warming counts, the EU needs to reinforce its climate policies, in particular regarding the non-CO_2 GHG emissions (e.g. methane, sulphur hexafluoride, etc.) which drive more than 30% of today's global warming.[10]

Of course, final EGD plans and their proper implementation will heavily depend on EU countries agreeing to make it happen. Here again bold determination by decision-makers will be required to transform encouraging intentions into concrete action.

To help drive these changes, Ursula von der Leyen recognises that the transition must be a just transition for all. To that end, the EC proposed in January 2020 the creation of a "just transition fund" of 100 billion euro for the period 2021–30, an amount that has been increased to an estimated 150 billion euro following the EC adoption of its new 2021–2027 budget proposal and of a 750 billion euro virus recovery fund.[11] This fund should provide support to those regions whose economy is the most dependent on carbon industries. State aid rules should be amended to enable governments to help those enterprises that are the most affected by the Transition. In total, the investment needed to meet carbon neutrality by 2050 would amount to 260–300 billion euro a year. 25% of the EU budget would have to be devoted to the Transition and this money would have to be both public and private.

EU legislation requires that each Member State drafts a 10 year National Energy and Climate Plan (NECP), setting out how to reach its national targets. The EU Member States are busy preparing national plans to revamp their current climate policies on the basis of the EGD and its climate neutrality objective. In 2019, the EC analysed the draft NECPs and found that a large majority of them were not geared to this new objective.

France has taken an innovative approach by asking 150 citizens, drawn by lot and gathered in a socalled Climate Convention, to come up with proposals for the future climate legislation and a similar initiative is underway in the UK Some political parties and civil society groupings have also made very progressive proposals at the national level to set up a new social and environmental contract with society. They are motivated by the gains that can be achieved by caring simultaneously for environmental and social issues through integrated packages of measures. In France, 19 organisations have come up with an "environmental and social pact"[12] made up of 66 proposals "to give everyone the power to live".[13] The French trade union,

---

[10]https://www.journalgeneraldeleurope.org/en/2019/11/20/dr-georgios-amanatidis/. Accessed 31 May 2020.

[11]https://www.euractiv.com/section/climate-environment/news/number-crunching-on-expanded-just-transition-fund-continues/. Accessed 31 May 2020.

[12]https://www.cfdt.fr/upload/docs/application/pdf/2019-03/pacte_pouvoir_de_vivre.pdf (French). Accessed 31 May 2020.

[13]Name given to the 66 proposals.

the Democratic Confederation of Labour (CFDT) and its general secretary, Laurent Berger, originated the pact that brings together diverse associations (environmental, educational, solidarity), unions, and mutual societies. The bottom line of the pact is the mutual reinforcement of social and environmental issues. The pact calls for more and better power sharing in a renewed democratic process, with concrete proposals to give a voice to the most disadvantaged and to better take into account the work of associations and unions. This contribution proposes another model of development that makes the Transition the driver of new social progress.

In the months and years to come, the pressure on governments to beef up their climate plans will intensify. Reluctant governments could possibly take advantage of having the EC take the lead with its EGD project, to relieve part of the negative national pressure exerted on them. The adoption of more ambitious measures at the national level may be facilitated if part of the responsibility can be passed up to the EU level. The full impact of the COVID-19 crisis may take some years to be understood but it is clear that it will affect the timetable and priority of EC proposals as resources and capacity are directed towards the urgent health and economic issues. Nevertheless, the fact that EU environment ministers have resolved to keep the EGD as central to a resilient recovery after COVID-19 is an encouraging sign.[14]

The multiplication of plans by national and local authorities, combined with their generally higher level of ambition compared to current policies (Finland to become carbon neutral by 2035, Oslo by 2030[15]) is the sign of a change in political priorities-at last! The general public rightly perceives that the local level is the most suited to take concrete measures pertaining to ambitious climate plans decided at a higher institutional level.

Based on the blueprints of the current plans, it seems that a large majority of them do not focus enough on how the policy measures have to be decided, endorsed and implemented to maximise the chance of success. Instead, the plans essentially concentrate on the "technical measures" for reducing carbon emissions through, for instance, the insulation of houses or the electrification of the transport sector. So far, few of them address the more fundamental issue of the orientation of the economic process. Another caveat concerns the practical implementation of the plans, and especially of the EU EGD. It is premature to speculate about the final level of ambition and the likelihood of these plans achieving agreement. Indeed, they risk being watered down in the long and complex decision-making procedures leading to their final implementation, and that is especially true for EU plans.

Climate neutrality entails a radical transformation of our production and consumption patterns. Are current plans enough ambitious to reach this goal in at least 30 years? The achievement of the plans' goals might also face significant obstacles for one main reason: the failure to integrate a convincing set of measures to adapt lifestyles to tomorrow's new reality.

---

[14]https://www.euractiv.com/section/energy-environment/news/france-germany-join-group-of-10-eu-countries-calling-for-green-recovery/. Accessed 31 May 2020.

[15]https://www.euractiv.com/section/climate-environment/interview/oslo-mayor-heres-how-we-plan-to-become-a-carbon-neutral-city/. Accessed 31 May 2020.

Climate Protection as the Central Policy Goal?

In a world facing threats from all directions which should be given the most attention by politicians and the general public? Should it be public health, climate change or the overconsumption of natural resources? Or something else?

The world has undoubtedly been changed by COVID-19 which, for a time, replaced climate change in the media as the most worrying issue faced by humankind. Although our responses to global crises have been re-shaped by the pandemic a growing number of persons, whatever their origin or degree of expertise, argue that if there were a hierarchy of the most pressing challenges, climate change would have to be put at the top. Such a claim has been endorsed by environmentalists for many years, arguing that climate change is the fundamental existential crisis for the human race.

Even if there is a relative consensus on the prominence of the climate challenge, there must be a careful reflection on the implications of giving priority to its resolution through the political process. Some authors, such as Richard Heinberg,[16] consider that the shift caused by climate change dominating political concerns is far too restrictive and prevents other very important environmental issues, such as biodiversity loss and overpopulation being addressed. He believes that our core environmental problem is not climate change but the overshoot of the earth's carrying capacity, of which global warming is a major symptom. Overshoot being a systemic issue, its resolution should bring benefits to all its environmental symptoms including climate change. He feels addressing climate change as a unique environmental symptom will not cure the root causes of the malfunctioning of our global ecosystem.

It is however encouraging that climate change having progressively found its place in the speeches and programmes of decision makers, was frequently referenced as a *sina qua non* in the planning of responses to the pandemic. Indeed, some actions towards sustainable living already undertaken by the public in response to the climate crisis were reinforced and extended by the pandemic. Even if there might be good reasons to believe that there are other equally important issues with the same or an even higher degree of priority, it seems quite complicated if not counter-productive, to bring about a change of political focus. It already took many years for governments to start focusing on climate change. If the new political priority was changed to reducing the overshoot then it is likely that this more complex understanding of global sustainability would require further explanation to the general public.

There is a further difficulty about promoting the overshoot into the number one position of concern. That would immediately identify excessive, wasteful and unequal consumption as the prime target of legislative effort and corresponding state-led public adjustment. This is probably the hardest and most threatening challenge imaginable as answering it involves a radical reassessment of everyone's lifestyle and expectations. It is therefore sensible and practical to continue mobilising societal forces by waving the climate flag but to do so in an integrated manner with other

[16] Heinberg (2017).

pressing issues. A key action point of the manifesto refers to the manner of defining and carrying out integrated approaches that do not disqualify minor policy objectives. In this respect, the line taken by the EGD is to be praised. Whilst pursuing an overall climate goal, i.e. putting Europe on track to reach net-zero global warming emissions by 2050, the EGD is made up of a diverse set of policy measures that find their full legitimacy through their relation with others.

Now that politicians have, at last, started to get worried about climate change, governments must make sure that anything that is decided to counter it must also, whenever possible, support other important policy goals which promote sustainability.

Acting for the Climate Is Acting for Society as a Whole

Europeans had become risk averse then risk appeared from nowhere in the shape of COVID-19. Until then climate change, with its potentially devastating impact, had not yet led to the eureka moment in terms of social awareness. Will this change?

In our society marked by individualism, acting in defence of a cause has increasingly been considered as something from the past. The revival of civil movements animated by considerable popular engagement in the last 2–3 years is a statement to the contrary.

At the start of a climate protection initiative there is, however, considerable political and social resistance. Resistance can be against change and even against positive change, especially when divergent political forces or lobbies pull the strings from behind the scene. Resistance from people and by people is of a different nature. People are against measures that they feel, wrongly or rightly, limit their freedom. Even when faced with an immediate threat to life a minority of protesters in many countries, but notably the USA, argued their civil liberties were more important than observing restrictions to counter the coronavirus pandemic. As the poet W. H. Auden wrote in his poem *The Age of Anxiety* (1947): "We would rather be ruined than changed." Naomi Klein does not hesitate to speak about a "deep anger and rage that some people can express in front of the idea that someone wants to impose limits".[17] To illustrate the inconsistent and senseless behaviour of people in thrall to populist political influence she often refers to the plastic Trump straws (with the logo Trump 2020) that were produced in reaction to the environmentally more friendly paper straws distributed by the Democrat camp. These plastic straws were massively purchased by Trump voters who displayed with pride their non-environmentally sensitive behaviour. On climate, as with so many other sectoral issues, the political polarisation of debates to the extreme constitutes a substantial blocking element to further progress on more substantial lifestyle changes.

---

[17] Klein (2017).

Part of the resistance also comes from another idea that ought to be challenged: it is not true that the sum of environmental policies are necessarily and always giving a negative result in terms of quality of life and future prospects, and that there is no other solution to this equation. An inspired and energetic leadership must have the strength to say that caring for the climate can be good for the economy and society when the measures are well designed. Sustainability as a whole is not necessarily about compromise and sacrifice. There are many good stories to tell people in relation to the synergetic effect of climate-related measures. Think of the numerous green jobs that are created by the renewable energy industry, the improved affordability of goods and services when they are co-created by local communities, the health benefits of cycling or eating less processed food.

Even if there is a progressive convergence of views that climate protection can go hand in hand with the promotion of other social, economic and environmental goals, advancing this idea will still require much effort. Governments have therefore to understand that if they decide to address climate change as one of their key priorities they need to do so by constantly reminding stakeholders of the potential winwin effects on other policies. Carrying out policies in a holistic manner requires appropriate impact assessment that takes due account of differentiated impacts for different policy areas (biodiversity, agriculture etc.) and social groups (by income, age etc.). The results of the impact assessment will need to be taken into consideration to calibrate the different policy and support measures necessary to reach the central climate objective and the equally important just transition.

To support holistic approaches, there is a further need to tailor communication to make the connections between the policy goals clear to every stakeholder and thereby illustrate that they are, indeed, coowners of a joint enterprise.

One of the key reasons for acting holistically when addressing climate change is that many of the pressing issues faced by today's society are accentuated by climate change. In its October 2017 opinion on climate justice,[18] the European Economic and Social Committee (EESC) stressed that the most vulnerable and poorest sections of society are often most affected by the impact of climate change, although they bear the least responsibility for the emissions which have led to the climate crisis. The findings of the EESC are corroborated by many national or regional studies that all draw the same conclusions.[19]

If the vulnerable are negatively affected then environmental measures will be felt as an injustice, a further imposition on those least able to adjust. People on low incomes are forced through the price mechanism to consume less energy or water or forgo other things they value. Sometimes this can be done creatively but frequently it can result in a reduction in quality of life or comfort. The prohibition of wood stoves, increased taxation on diesel fuel or the creation of low emission zones in city centres are all justified from an environmental point of view but pose real difficulties to the

[18]https://www.eesc.europa.eu/en/our-work/opinions-information-reports/opinions/climate-justice-own-initiative-opinion. Accessed 1 June 2020.

[19]Durability and poverty, 10th Biennial Report of the Belgian Service for the Fight Against Poverty, Poverty and Social Exclusion, 11 December 2018, https://www.luttepauvrete.be/ (in French).

ones who cannot afford the sometimes costly alternatives. Electric vehicles remain out of reach for poorer households and for a large majority of goods (for instance low cost clothing) the affordable options are very often not environmentally friendly.

Finding ways to address climate change and other related issues together is a prerequisite for public acceptance but can also enhance the cost-efficiency and the political legitimacy of proposed measures. Our conviction is indeed that the resolution of climate change will require profound transformation in many aspects of our social, economic and cultural fabric. When applied in isolation, climate policies can, indeed, lead to extra costs for many. To increase their level of acceptance, they must be combined with other policies that have a counter-balancing or synergistic effect. One of the most obvious illustrations of the possible win-win outcome relates to simultaneously addressing inequalities and climate change. Environmental taxes cannot be understood and accepted without giving opportunities to low income households to reduce or even annul their tax liability, itself a driver for reducing inequalities. Similar positive combinations could be reached by pairing climate change actions with actions on biodiversity, on improving the sustainability of the food system or on reducing the tensions felt by national producers resulting from the imports of cheap goods from external markets. The door has been opened for such measures by legitimate concerns about globalisation arising from the coronavirus pandemic.

Jointly approaching climate and biodiversity issues makes a lot of sense. The crisis of biodiversity looks similar to that of climate change in many respects and could lead to equally disruptive situations in the medium term. This was highlighted in the May 2019 report of the Intergovernmental Science-Policy Platform on Biodiversity and Ecosystem Services (IPBES)—often described as the equivalent of the IPCC on biodiversity. In the IPBES report, climate change is pointed to as the third most significant driver of biodiversity loss but its impact could surpass one of the other drivers in the coming decades. The Report also found that global goals for sustainably using nature cannot be met by current trajectories, and goals for 2030 and beyond may only be achieved through truly transformative changes across economic, social, environmental, political and technological factors. Other key indirect drivers include increased population and consumption per capita but also, more critically, issues of governance and accountability that are pivotal for achieving progress in the mitigation of biodiversity loss.

Nor should other, apparently less relevant issues be excluded from incorporation in a holistic policy approach to the wider climate picture. Women's rights, Fair Trade and animal welfare all have a part to play, along with many more. The goal should be to aim at a fully integrative approach and assess the possible synergies that can be achieved. What is at stake is not only to put together different policy goals but to aim at their full convergence and conciliation within a supporting socio-economic framework. The latter requires smart governance engineering and is essential to guarantee that all the sectoral actors defend the climate goal with the same vigour.

The goal of an all-inclusive climate policy is paramount. Positive action for the climate can lead to many constructive multipliers for society, so why wait?

People Are Becoming Concerned—And Active

Young people are shaking the foundations of society. In striking for the climate, they do not defend their own interest but everyone's interests. Not everyone is going to think the same way as climate protesters overnight, but many will start to think.

At the heart of the impressive surge in climate protests which emerged in mid-2018, there is fear - a fear that endures and progressively embeds itself in the minds of people and affects the way they live. Fear is paralyzing, it breeds panic and desperation, it blocks reactions and engenders negative, counterproductive feelings. Faced with the big, first-of-a-kind threat that is climate change, there would be many good reasons to succumb and do nothing. But young people have not succumbed. They have shown the way with positive action.

In contrast, do nothing is the preferred option of those supporting the current economic model based on harsh competition, globalised trading and infinite consumption. In line with the There Is No Alternative slogan (TINA), neoliberalism does not propose any vision of the future in which it would be desirable to change. There are more and more dissonant voices expressing doubts about the validity of today's economic model. No one seems to be disposed any more to say loudly that the situation would improve if one would continue to deregulate or privatise or relaunch domestic consumption. Neoliberalism has come to its terminal point and has lost its capacity to steer people towards a positive project.

The spreading despair is the consequence of the lack of credible alternatives proposed by the establishment. In the last 20–30 years, there has been no single, credible proposal to tackle the root causes of climate change from the current economic system, which remains largely based on the doctrine of make/use/dispose/replace.

First in line to be held accountable are the politicians, as they are the ones in control of the policymaking and of the overall regulation. We have now reached the point where the inaction of governments has become unjustifiable and irresponsible. Politicians tend to affirm very boldly that they have understood the desire for action by an ever growing share of the population but they generally do nothing. In COP meetings, the speeches of heads of states are full of responsibility and compassion, but the action does not follow.

People aspire to another model of society. The demand for a change will intensify in the months and years to come, even if measurable behavioural changes are still to come. An interesting survey undertaken in France by the Observatory Society and Consumption[20] shows that a growing number of citizens are seduced by a form of environmental utopia. Out of three models of society, more than 50% of respondents (55%) express a preference for lifestyles that are based on "better" consumption, proximity, quality, social links and slowness. The most disqualified model (only 19% of positive responses) is the techno-liberal one that is the closest to today's

[20] L'Observatoire des perspectives utopiques, October 2019, see http://lobsoco.com/lobservatoire-des-perspectives-utopiques/. Accessed 1 June 2020.

model as it favours a hyper individualistic society with a strong growth driven by science, technology and trans-humanism.

With the climate risk materialising more quickly than anticipated, it is no surprise that the first, massive reaction to the status quo has come from the Millennials. Climate protection is becoming the field of a new, assertive form of expression of young people against society in general and the older generations in particular. Until recently, adults were almost entirely determining the education of children, with assumptions of experience, knowledge and rationale supported by their apparent maturity. In challenging the choices and ways of living of their parents, young people are forcing the society of adults to a self-examination. Forced to re-evaluate their way of living and their material consumption habits, adults appear mute in front of empowered youngsters. It is not easy to hear from those who are supposed to follow you that there cannot be any progress anymore and that it is just a question of surviving. One has to rejoice that more and more young people are breaking free from indifference, ignorance and submission. Young people leaving school to teach adults is a picture that will be hard to forget. It is actually a strong encouragement for driving the collective effort to protect the climate with rejuvenated enthusiasm.

The demonstrations that were organised in hundreds of cities from mid-2018 have also attracted a larger and larger amount of older people, showing that a growing number in society is asking for more decisive and immediate action to combat climate change. Over time, the climate marches have gained more and more recognition and popularity, attracting up to 4 million people on 20 September 2019 in more than 100 countries. The success of climate protests symbolises a step change in the way climate change is perceived culturally and internationally.

However, and this is the prime reason for this manifesto, the political response to this massive, international movement has so far been grossly insufficient. The expectation of people regarding the nature of the role to be played by governments has been analysed in a recent survey by Viavoice in France.[21] Respondents primarily trust NGOs and associations (55%) as well as citizens (48%) to fight climate change. States and governments, to the contrary, are deemed insufficiently active. The recognition of the important role of civil power does not imply that people dismiss the potential of politics. To the contrary, and quite surprisingly, 61% call for "a much more authoritarian role by the state", which is asked to impose "constraining rules". Only 32% favour incentivisation from the bottom. This demand for governmental authority is principally coming from the supporters of the left (71%), with right-wing sympathisers being divided on its relevance (54%). As far as could be generalised, a positive lesson of the survey is that should governments aim to regain credit and public support, they will be expected to carry out truly transformational action for the climate, and preferably with some form of command and control.

One of the reasons for governmental inaction is related to the fact that, despite its success, the international movement has not yet succeeded in attracting all sectors of society. If the enduring engagement of young protesters has generally been hailed,

[21] Libération, 21–22 September 2019, pp. 12–13.

it has also been mocked by minority political and social groups. The main misunderstanding comes from the low income, less educated households who have been globally absent from the marches. Doubts are rising regarding the further expansion of the movement beyond its strong youth and climate activists' base.

There are many reasons why people are not united in defending the cause of a vigorous fight against climate change. The world population is little informed about the causes and consequences of anthropogenic climate change. Improving the quality of information might help to strengthen mobilisation. However, it almost certainly will not suffice as there is no paucity of reports, op-eds, scientific articles, or resolutions and yet we continue sleepwalking.

Developing the sense of a world community that is sharing one unique planet would be a great booster. But again, this is very hard to achieve. The philosopher Francis Wolff[22] stresses that we, inhabitants of Mother Earth, have never been so close to the idea that we are forming a single, united humanity. A truly united humanity would be better placed to deal with the multiple, global risks with which we are increasingly confronted: climate change certainly, but also terrorism, cybercrime and economic crises. There are however, according to Wolff, opposite forces that erode the idea that we are forming a human community. He deplores in particular the identitarian isolationism that affects many regions, sometimes created by new forms of religious radicalism. Undertaking joint solutions to environmental challenges at planet level will require equally important efforts to safeguard universalist values and the coexistence of cultures. Preserving global human values across countries will be the strongest defence when facing the challenges of severe natural catastrophes or large waves of migration.

The fact that all countries will be affected by climate change but that the nature and intensity of impacts will largely differ from one region to another, is an additional difficulty towards creating a common and shared vision. Fostering a more connected international human community is also very challenging given the global spectrum of economic development. Huge differences will continue to exist between developing and developed states in the coming decades. According to an Organisation for Economic Co-operation and Development (OECD) report,[23] large inter-country differences in living standards will persist until at least 2060; the living standards in China, India and other emerging countries will still be 25–60% lower than those of leading countries in 2060. Even within countries, different social groups will perceive the threat of climate change through widely different lenses. The holding of climate demonstrations on the same day across continents is, at the very least, a sign of an emerging international consciousness - but this is just the beginning.

Beyond asking governments to act, there is the even more sensitive question of people's actual engagement to contribute through a change of lifestyle. According

[22] Wolff (2019).

[23] Looking to 2060: a global vision of long-term growth, OECD report, November 2012.

to a 2019 survey carried out by the EC (called Eurobarometer),[24] 93% of EU citizens see climate change as a serious problem, with 79% seeing it as a very serious problem. 92% of respondents agree that greenhouse gas emissions should be reduced to a minimum while offsetting the remaining emissions, in order to make the EU economy climate-neutral by 2050. A very large majority of Europeans are in favour of bold actions by their governments. 92% of respondents think it is important their national government sets ambitious targets to increase the amount of renewable energy used and 89% believe governments should provide support for improving energy efficiency by 2030. 84% believe that more public financial support should be given to the transition to clean energies, even if it means reducing subsidies to fossil fuels. But a gap continues to exist between the high levels of awareness and demand for political action expressed in these surveys on the one hand, and low personal engagement, on the other hand. Climate marches in Europe still only bring together a small minority of all those who place climate disruption at the top of their concerns. And an even more limited number of people are actually changing their lifestyles - how they eat, how they travel, how they insulate their homes (or do not), or what or how much they buy.

Why is getting engaged in positive climate action so difficult? Again, there are many different kinds of reasons explaining the relatively slow progress. Not everyone is well informed or has the resources to choose options for sustainable consumption. Not everyone is able to resist the pressures of our consumerist society to prioritise style, fashion, novelty and the trend to regard shopping for all sorts of goods and services as a self-affirming and enriching experience. People who want to make positive gestures for the climate have to be clear about what it means in terms of changing their way of living. Banning plastic from one's consumption is far from being enough.

There is also the question of who takes the first step and whether the 'burden' is being shared equally. The harsh personal and corporate restrictions that were introduced in response to the COVID-19 crisis involved a far greater lifestyle adjustment than anything envisaged for combatting the climate emergency. For the most part, they were widely and effectively observed, partly because they were seen as necessary, partly because they were likely to be short-term and partly because they were applied equally across society. The adjustments in lifestyle to help the climate are equally necessary but cannot be seen as temporary. However, they can be introduced gradually and in an equitable way so that, as with COVID-19, they become the 'new normal'. A similar sense of "We are all in this together" can support the joint public effort which will be required.

Experts agree that meeting even the modest proposals in the Paris Agreement will imply a fundamental reorganisation of our ways of producing and consuming and that it requires the commitment of all to reduce our carbon emissions. Everyone needs to adjust to doing things for the climate without immediately being able to see their impact. It is psychologically difficult in an age of growing individualism as it requires

---

[24]https://ec.europa.eu/clima/sites/clima/files/support/docs/report_2019_en.pdf. Accessed June 1 2020.

reorienting thinking towards the community, global wellbeing and future generations. Effective climate action requires subordinating self and advancing wide collective interests. It is definitely something that is not easy for the majority of citizens living in democratic states who regard their personal liberty and freedom to choose as unlimited. The inability to realistically picture a different lifestyle (perception that comfort and wellbeing would be reduced) and the impression that some would make efforts while others would continue "enjoying" impose that decision makers propose an attractive set of enablers for the lifestyle change.

Placing greater value on the benefits of climate protection and sharing a framework of universal values are essential in diversifying, expanding and consolidating the newly created grassroots movements. We all need to find our place in the climate movement and make our voice heard and our actions count.

Civil Society in Action

A 15 year old girl goes on strike from her school to highlight a cause that should have been the main focus of political and social action since before she was born. How does this story reflect on our priorities?

The figure that best represents the massive popular demand for bolder political action is Greta Thunberg, now a 17-year old Swedish girl. Acting on her own conviction and initiative in August 2018 she began to spend her school days outside the Swedish Parliament holding up a sign saying "School strike for climate". Her determined charisma has led thousands of youngsters to act similarly in their own country. In October 2019, at the official opening of the 2019–2020 academic year, the University of Mons (Belgium) delivered their doctoral diploma and insignia to Greta Thunberg alongside British economist Lord Nicholas Stern and former French minister Nicolas Hulot. In the face of criticisms by some, the Rector of the university, Philippe Dubois, argued that "Greta Thunberg contributes to raising awareness of the imminent threat of climate change to society. We have seen the hope she inspires. In choosing her, we recognise her role as ambassador of climate consciousness".

She is not alone; other charismatic figures picked up the baton, for example Anuna De Wever and Adelaïde Charlier in Belgium, Luisa Neubauer in Germany, Hilda Flavia Nakabuye in Uganda and Hiroto Inoue in Japan and in 2017 Jamie Margolin in the USA preceded her in founding the Zero Hour organisation at the age of 15—all are defending the cause of the climate in their own country.

The emergence of articulate young people to defend causes that apply across society, is an innovation. Major societal claims are no longer the exclusive responsibility of political parties, unions, NGOs, other kinds of representative organisations or adults. The role of social media in the emergence of these young individuals and the crystallisation of this new 'climatic' demand is enormous. Organic by its very nature, this process has not required any prior planning or structuring, hence the marginalising of traditional intermediate bodies.

Different movements were created in the wake of the civic shake-up instigated by Greta Thunberg. In Belgium, Anuna De Wever and Kyra Gantois created the Youth for Climate movement[25] in January 2019, as a response to the action of Greta Thunberg. Very quickly, Youth for Climate started becoming active in other European countries such as France, The Netherlands and Switzerland. To pressurise the EU to recognise and promote the demands of young climate activists, Generation Climate Europe (GCE)[26] was created as the first environmental coalition of the main youth-led organisations and networks active at the European level.

The majority of the civic efforts to coordinate forces are post-September 2018 but there were also other climate movements created before the emergence of Greta Thunberg. Jamie Margolin founded Zero Hour[27] with the aim of organising a climate March at the international level. On 21 July 2018, thousands of young people crowded into the streets of 26 cities. The Sunrise Movement,[28] created in April 2017, is a US youth movement to make climate change an urgent priority across America and elect people who stand up for the well-being of all citizens. The resistance to climate unfriendly projects has existed well before 2017 but the many social groups which opposed a variety of potentially damaging projects never gained the huge visibility that Greta Thunberg was able to secure through her name and activity.

The media coverage around the youngsters from the Iakota community, a Sioux tribe, who opposed to the construction of a pipeline in the North Dakota, has been important but has never received the kind of international attention that Greta Thunberg received through her sitting in front of the Swedish Parliament. The success of Greta Thunberg contributed to the shaping of a strong international movement, rallying the many climate protests that occurred at a national or regional level.

Each of the national climate leaders use their own personal values and temperament to mobilise young people and address politicians. Unlike Greta Thunberg, who committed herself not to travel by air, other leaders are keeping the flexibility to do so whenever deemed necessary. The national context sometimes tends to concentrate attention on particular aspects of the struggle. Consumerism is, for example, one of the angles through which Jamie Margolin orients her personal action. The personal decision of Greta Thunberg to not use planes, had a very large impact in Sweden with the launch of the flygskam or "flight shame" campaign. In the period January to October 2019, there was an 8% drop in passenger numbers on domestic flights and a 3% dip for international journeys. Recent trends seem to indicate that passenger numbers are steadily dropping on domestic and short-haul routes within Europe. Germany has seen a slump in domestic air travel which has mirrored the decline in Sweden, The German rail operator Deutsche Bahn has reported record passenger numbers whereas the ADV industry group reported a 12% decline in commercial aircraft passenger numbers for November 2019, the fourth straight monthly decline.

[25]https://youthforclimate.be/. Accessed 1 June 2020.

[26]https://gceurope.org/#gce. Accessed 1 June 2020.

[27]http://thisiszerohour.org/. Accessed 1 June 2020.

[28]https://www.sunrisemovement.org/. Accessed 1 June 2020.

Beyond their personal characteristics, the majority of young climate leaders are showing similarities in the way they present themselves, displaying a high level of self-confidence, no hesitation to name and shame, and speaking directly without filtering the message. Also refreshing is the healthy irreverence with which young people address their elders. Another striking element is the fact that all the figures who have recently emerged to found or lead direct action movements are young women. So far there have not been any male leaders heading these climate protests on behalf of the planet, quite the opposite to what happened in the civil unrest in France in May 1968 when most of the leaders were men!

The youngsters who have courageously decided to make personal sacrifices in defence of the climate have nevertheless not yet won the battle for public opinion. Even with the massive and generally supportive media coverage, young activists have not been able to convince the rest of the population of the necessity to mobilise 100%. The best indication of this is that only 10% of votes were given to specifically green candidates in the European elections of 26 May 2019. Whether this truly showed that European citizens did not place the climate at the top of their concerns is uncertain but with this kind of weak signal, politicians in Europe will not feel pressed or obliged to perform the necessary environmental restructuring of the economy. The new European Commission led by Ursula von der Leyen only includes one Member coming from a green party, the Lithuanian Member Virginijus Sinkevicius.

In a Brussels debate with EU Ministers on 5 March 2020,[29] Greta Thunberg was greeted by the Danish energy minister Dan Jørgensen in the following terms: "It sometimes takes a child to challenge the status quo and articulate an uncomfortable truth," citing Danish author Hans Christian Andersen in "The Emperor's New Clothes". In spite of this one good political response, too many remained mute. The demands of young people to listen to scientists and to forget about the recipes of the past have not yet had the desired outcome to press politicians to act. In the past few years, politicians have had multiple opportunities to take on board the message of scientists. They can no longer claim ignorance about the nature of the threat and the collective obligation to reduce carbon emissions. But they still find the charge of being responsible for disruptive effects in our present economic system too uncomfortable to act more decisively on this acquired knowledge.

What else can be done to encourage governments to assume their responsibility and take action? The very gloomy outcome of the COP25, presided over by Chile and organised in Madrid, has reinforced the conviction of the pro-climate activists that all means must be used to force governments to act. Increasingly, people will use many and more varied means to conduct the climate fight.

An increasing number of citizens and NGOs have started to use legal frameworks to hold their governments accountable for failing to protect people and the planet. Courts have in general taken decisions which recognise that governments are failing to protect human rights due to their climate inaction. In October 2019, the Berlin Administrative Court made a decision on the first ever climate litigation case against

[29]https://www.euractiv.com/section/climate-environment/news/thunberg-warns-eu-ministers-about-worlds-declining-carbon-budget/. Accessed 1 June 2020.

the German government. The complaint was filed by three farming families and Greenpeace Germany against the German government for failing to meet its 2020 climate target. The case was dismissed but the Court recognised the right to a safe climate. In December 2019, the highest court in the Netherlands upheld a ruling requiring the government to slash greenhouse gas emissions by at least 25% of 1990 levels by the end of 2020. In February 2020, the British Court of appeal ruled illegal the plans for a third runway at Heathrow airport on the grounds that ministers did not adequately take into account the previous government's decision to set a target in law of net zero emissions by 2050. The court's ruling has been interpreted as the first major ruling in the world to be based on the Paris climate agreement. Brought by local residents, councils, the mayor of London, environmental groups including Friends of the Earth and Greenpeace, and the legal charity Plan B, the case was first dismissed in the high court in May 2019. There were comparable cases in Ireland, Italy, and other European countries, just to name a few.

Tomorrow's pressure from civil society will be a mix of conventional protest and legal challenges. The latter will send stronger and stronger signals to governments and investors that a radical change of direction is necessary.

Act Two of the pro-climate movement will be something else, more varied and possibly tougher than mere street protest or court actions—after all many see the future of humanity as worth fighting for.

The Emergence of Civil Disobedience

Civil disobedience is being resorted to more and more frequently and the risk of social polarisation is looming. Could there be agreement across the whole of society to act on behalf of the climate?

The inaction of governments is gradually bringing the most engaged citizens to consider the idea of disobedience, of behaving at the edge of the law or beyond it. Civil disobedience is seen as a way to reclaim a voice that many feel they have been deprived of.

The sequence of climate mobilisations since the end of 2018 revealed their usefulness to force home messages that are still too largely unknown but also, regrettably, their lack of impact on the political agenda and personal action. There is plenty of evidence that many of those participating in marches have not changed the way they live and consume. Attending public protests is definitely a positive gesture but it remains a first, small step in the right direction. The American philosopher Michael Walzer, author of the manifesto of civil action "Political Action: a practical guide to movement politics",[30] regrets that the large number of people who demonstrate in the street quickly lose contact with the organising movement once they return back home.[31] He is a strong supporter of the international action led by Greta Thunberg

[30] Walzer (1971).

[31] Interview in the Belgian newspaper Le Soir, 29 October 2019, p. 18.

but thinks that it has to be backed by actions at the local level: "One must establish a cross border organisation relying on national cells, which themselves have to organise mobilisations and intensify the pressure on the local governments".

A large number of those who have participated in the climate mobilisations in 2019 share the view that it is time to change the channels of protest. Civil disobedience is increasingly becoming an acceptable option to pursue mobilisation through means other than the marches. A study by the French collective, Quantité Critique, showed that a very large majority of those participating in the climate demonstrations are supportive of more dramatic civil disobedience action, such as blocking coal mines or the operation of polluting industrial plants. Coincidently, other sectoral conflicts, especially the ones relating to the basic rights of people (salary, pension etc.), are leading more and more often to cases of violence, evidencing the great despair of part of the population and the lack of response at a higher level. That a peaceful pro-climate movement decides to disobey or that socially disadvantaged people feel authorised to use violence to defend their existing rights, are acute signs of a suffering democracy.

Nevertheless, climate defenders who are tending to radicalise remain committed to non-violence. Civil disobedience actions should, in their eyes, be principally blockage actions. In the same way as a trade union's action would picket work places, pro-climate civil disobedience actions target the blockage of commercial centres or of main roads in front of buildings symbolising the political power. Though peaceful, these actions are on the edge of legality. The international movement of civil disobedience Extinction Rebellion (XR) is the most visible sign of the changing mind set of protesters. XR defines itself as an international movement that uses non-violent civil disobedience in an attempt to halt mass extinction and minimise the risk of social collapse. Its very existence is evidence that the patience of part of the public is just as finite as the natural resources available on the planet earth are. It is also highlighting the feeling of powerlessness that progressively permeates the most active protesters. Time is running out and action must start now, this is the bold message of XR. No doubt this message will be recalled in many different ways in the weeks, months and years to come if inaction continues to be the preferred option of governments, as demonstrated by the poor result of the COP25.

The growing public support for civil disobedience also highlights the conflictual dimension that crosses the environmental movements in many different countries. The truth is that the progressive and consensual process of deliberation between the decision-makers and civil society has not delivered the kind of transformational changes that are required for addressing the climate problem. It is therefore no surprise that more and more voices are calling for a radical change in the way of influencing the decisions at the top. In the near future governments will have to prove their capacity to propose bold and convincing steps to operationalise the Transition, taking possibly inspiration from the European Green Deal. Not doing so could mean that more extreme movements than XR will emerge and exert more radical means of action in the public space. The toolbox of protest has been opened and will be thoroughly explored. More extreme and dramatic forms of protest, if used, will lead to more social antagonism and more hurdles in uniting the whole of society behind the

Transition. In the longer term, there is some risk that climate issues, if not adequately handled by governments, will fuel scenarios which may lead to democracy's exit.

Even if there remains a gap between the number of people who accept the idea of civil disobedience and the still small minority participating in such actions, the steady growth of this group of activists over time seems inevitable. The work by Quantité Critique point out that the ones who have already practiced the blocking of infrastructure and the ones who are ready to do so, tend to consider that the Transition cannot be performed within the capitalist system of today. They are also the ones who are the most inclined to modify their way of living to reduce their personal footprint.

In Canada, indigenous people have been blocking rail services from the beginning of 2020 to protest against oil and gas projects on their own land. In the last week of February, the Vancouver-based company Teck Resources announced that it was cancelling its Frontier megaproject that was aimed to become the largest bituminous sand project in Canada. The company said that their decision has been caused by the pressure and blockades of the North British Columbia Wet'suwet'en tribe. This victory is symbolic of a change in the way activists proceed to get their cause heard. Diverse Canadian environmental NGOs have fought to oppose to similar projects but hitherto in vain.

In February 2020, about 1000 scientists published a call in the French newspaper Le Monde for "everyone to participate in civil disobedience action organised by the environmental movements, be they historical (Amis de la Terre, Attac, Confédération paysanne, Greenpeace etc.) or recently created (Action non violente COP21, Extinction Rebellion, Youth for Climate etc.). By acting individually, by coming together at the professional or local level, or by joining existing associations and movements, scope for action will appear to unlock practices and develop alternatives". The main motive of this large group of scientists is that "for decades, the successive [French] governments were unable to put in place strong and rapid action to confront the climate and environmental crisis whose urgency grows daily.[32]

It is, however, necessary to recognise that civil disobedience is a two-edged sword in several respects. By acting on the fringes of the law, the status and value of the legal principles that uphold effective governance are inevitably called into question. In addition, it is open to all types of groups to take to the streets to question the action of governments. The *gilets jaunes* in France are a good example. There are many other examples of street protest around the world and even, in some cases, of isolationist and nationalist protests. Even when faced with an existential crisis such as COVID-19, libertarian groups across America were quick to protest against the restrictive actions taken by federal and state authorities. The fact that a means of protest can be applied to many purposes does not invalidate its use but is a reminder that it should be used cautiously.

Even though organised civil disobedience does increasingly appear as a necessity, the resolution of climate challenges will require, sooner rather than later, a combination of all forces in society taking decisions in common. A movement such as XR

---

[32]Face à la crise écologique, la rébellion est nécessaire, in newspaper Le Monde, 21 February 2020, p. 22.

will have to show its ability to enter into political negotiations and to accept compromises if it wants to be part of the formulation of responses to the climate crisis and propose alternative social models that can be supported by a large part of society.

Symbolic actions in the public space, using humour, personalities and social media, could help to raise the awareness of all, if well designed and non-violent. But unless such actions are met by a positive response from governments then events could become damagingly unpredictable.

References

Amanatidis, G Briefing "European policies on climate and energy towards 2020, 2030 and 2050", PE 631.047, European Parliament, 2019.

Heinberg R, There's no app for that (2017) Post Carbon Institute, Santa Rosa.

Klein N. (2017) No is not enough, Allen Lane, London.

Klein N., (2019) On fire, Simon & Schuster, New York.

Pitron G, La Guerre des métaux rares (2018) Les Liens Qui Libèrent, Paris.

Rifkin J. (2019) The Green New Deal, St. Martin's Press, New York.

Walzer M. (1971). Quadrangle Books, Chicago.

Wolff F. (2019) Plaidoyer pour l'universel, Fayard, Paris.

Chapter 3
The Paris Agreement—Getting People's Buy-in Now!

Abstract The Paris Agreement does not speak to hearts and minds. It has to be translated in clear, meaningful targets that the general public can apply, day-by-day. To become mainstream, climate protection will require major efforts from governments, business and the general public. Informing, enabling and financing policies have to be in place to oil the wheels of a sustainable economy and engage the consumer. In short, climate decisions require smart and courageous governance that gives clear direction and provides opportunities to all actors of civil society. For this to happen, the process of making and taking decisions must be entirely reformed to favour ambition, collective engagement, practical implementation, and the fixing of limits for the public good, even if unpopular.

The general public needs targets, clear signals and concrete tips if it is to be persuaded to adopt carbon neutral approaches. The exact emphasis is problematic for policy-makers. Maximum temperature increase? A relative CO_2 reduction objective? A limited carbon budget per person? Do's and don'ts lists?

The Paris Agreement's central aim is to keep global temperature rise in this century no more than 2 °C above pre-industrial levels and to pursue efforts to further limit the temperature increase to 1.5 °C. Though citizens comfortably master anything relating to temperature and weather forecasting for their next weekend, the time scale of the Paris Agreement seems irrelevant to them. For some living in countries where the average temperature is lower compared to other parts of the world such a temperature increase can even be received positively. Mobilising people around this official goal can be either confusing or unexciting, hence the obligation of policy makers to translate Paris into more pragmatic objectives. The reality is that none of the climate goals set so far by policymakers have had any outstanding impact at the level of people's ordinary life.

E. Ponthieu, *The Climate Crisis, Democracy and Governance*,
SpringerBriefs in Climate Studies,
https://doi.org/10.1007/978-3-030-58127-5_3

CO_2 makes up just 0.04% of the Earth's atmosphere so encouraging the public to take seriously a relatively small reduction in this molecule can be seen as a challenging task. Initially the first set of goals established by the EU in 2009[1] were principally addressed to major stakeholders and industrial players in particular with, for instance, specific targets relating to the uptake of renewable energy and the improvement of energy efficiency.

There is an argument that had ambition—and impact—been greater and made more public, it would have stimulated a more creative, informed and educational public debate. By contrast, the measures taken by governments around the world to combat COVID-19 have had a dramatic effect on people's freedoms and finances and as a result have been the subject of intense analysis and deliberation. It is this level of public debate and awareness that will be necessary as a formative background to climate action.

Even though governments more recently started to feel the need for a wider stakeholder engagement in climate mitigation efforts, they kept using the same relative reduction objectives without attempting to translate these into more meaningful and concrete aims that refer to how and how much people consume. This reflects above all the difficulty to fix consistent ground-level targets that are both well understood and supported with a clear route map. The very existence of many different ways to reduce daily carbon emissions implies that there cannot be a single whole-encompassing objective, but a multitude of specific, sectoral objectives. Furthermore, each sector of activity contains substantial vested interests concerned with the way carbon mitigation efforts are spread over time and shared across stakeholders, hence a total absence of consensus on the hierarchy of measures to be put in place. Stakeholders harshly compete to transfer blame and responsibility to other parties to avoid being forced to pay a disproportionate share of the overall carbon reduction endeavour.

In creating a public impression that climate protection policies are not of the highest priority, governments have put themselves in a sort of schizophrenic situation where one of their most important political goals has modest societal acceptance and little foundation. In a majority of EU countries, the relative CO_2 emission reduction objective has been represented through a limited number of specific sectoral objectives. This is notably the case of the transport and energy sectors which have been specifically targeted by definite pieces of EU legislation. For instance, CO_2-based taxation of passenger cars is now well established across the EU, with 22 EU member states applying some form of CO_2 tax to the acquisition or ownership of cars.[2]

In general, there were few efforts made to ensure that other important sectors of activity such as the food or the tourism sectors, adapted to the Paris Agreement. Food

---

[1] 20 20 by 2020—Europe's climate change opportunity {COM(2008) 13 final} {COM(2008) 16 final} {COM(2008) 17 final} {COM(2008) 18 final} {COM(2008) 19 final}.

[2] CO_2-based motor vehicle taxes in the EU, study by ACEA (European Automobile Manufacturers Association), June 2019. By June 2019, the 6 countries that did not apply CO_2-based taxation were: Bulgaria, Estonia, Hungary, Lithuania, Poland and Slovakia.

accounts for over a quarter (26%) of global greenhouse gas emissions.[3] According to an international study published in 2018,[4] the tourism sector accounted for 8% of all anthropogenic emissions of CO_2 in 2013. In the medium term, when the covid crisis will be over, its carbon footprint is expected to increase significantly due to a lack of regulation.

In reality, there has been very little systematic alignment of sectoral policies with the goal of the Paris Agreement and the situation is pretty similar when one considers the implementation of the 17 Sustainable Development Goals (SDGs) pertaining to the 2030 Sustainable Development (SD) Agenda. In their Sustainable Development Report 2019,[5] Sustainable Development Solutions Network SDSN and Bertelsmann Stiftung show that many countries have not yet taken the critical steps to implement the 2030 SD Agenda, though it is a radical, equitable and achievable agenda. The report is a stark revelation of the huge gap between rhetoric and action. Out of 43 countries surveyed, including all G20 countries and countries with a population greater than 100 million, 33 countries have endorsed the SDGs in official statements since 1 January 2018. Yet, in only 18 of them, do central budget documents mention the SDGs. And it remains to be seen how this mention translates into concrete action.

Another international study reveals that a very large majority of countries have not translated their Nationally Determined Contribution (NDC) targets (to which they committed under the Paris Agreement) into national laws and policies.[6] Setting robust targets in national laws and policies is however crucial to the credibility of countries' commitments to the Paris Agreement. By October 2018, 157 of 197 Parties to the Agreement have set economy-wide emissions reduction targets in their NDCs,[7] but only 58 have done so within domestic laws or policies and only 16 of those are consistent with those set in the NDCs. Furthermore, more than 60% of economy-wide and sectoral targets in national laws and policies were set to 2020, while the target year for most NDCs is 2030.

Greta Thunberg disputed the relevance of setting national and regional relative carbon reduction objectives without having proper consideration for the international picture. In March 2020, in several meetings with EU officials, she insisted on policy-making being based on the amount of carbon dioxide that world nations are still allowed to emit before the rise in global temperatures exceeds the Paris goals. In the last report by the Intergovernmental Panel on Climate Change (IPCC), the world's remaining carbon budget—the amount of CO_2 that world nations can emit before climate change risks spiralling out of control—is estimated to be around 420 Giga

[3]Ritchie and Roser (2020).

[4]The carbon footprint of global tourism, Lenzen et al. (2018).

[5]Sustainable Development Report 2019—Transformations to Achieve the Sustainable Development Goals, SDSN and Bertelsmann Stiftung, 28 June 2019.

[6]Nachmany and Mangan (2018).

[7]The EU has submitted an NDC on behalf of its 27 Member States.

tonnes (GT) of CO_2.[8] If the nations of the world continue to emit carbon as at present, she said that "the remaining budget will be gone before the EU will even have a chance to deliver on its relative carbon reduction targets for 2030 and 2050".[9] She recognised that the carbon budget was in itself incomplete for proper policy-making, in particular as it lacks any reference to equity aspects and the responsibility of industrialised nations to more intensively reduce their emissions compared to poor countries.

Another track worthy of consideration by decision-makers is the allocation of personal carbon budgets (PCB) to EU citizens, as a way to raise awareness and incentivise people to reflect upon the climate impact of their actions. A PCB has the potential to be an important component of an ambitious national carbon reduction strategy. Citizens would receive an individual carbon emissions' budget for their personal use, which would correlate to the level of reduction aimed at national or regional level. Though PCB schemes have been intensively researched, they have not been so far implemented. To be practically operational, it is often proposed that PCB schemes only cover emissions under individual's direct personal control, such as household energy use (mainly electricity and gas), private transport (not including public transport) and aviation. In contrast, it would not include the carbon embedded in products and services purchased by the individual, such as the soap or the fruits bought in a supermarket. People could be allowed to buy additional emissions or sell their surplus credits in the personal carbon market. To be even more effective, PCB schemes should be integrated into an overall behavioural change programme to give a premium to those individuals making carbon savings.

The main reason why the purchase of non-energy and non-transport products cannot yet be part of a PCB scheme is the unavailability of a standardised carbon footprint methodology for products and services. The EC has developed, as part of its Action Plan on the circular economy and in consultation with sectoral stakeholders, pilots for measuring the environmental footprint of different categories of products and sectors[10] but there is still a long way to go before CO_2 codes of colour would appear on display besides the price of goods and services. The absence of objective carbon labelling is a major weakness for driving consumption towards more sustainability, and especially for countering the wrong beliefs regarding the "good" or "bad" products and services on the sustainability scale. For instance, what Europeans ate has the biggest impact on their footprint and contrary to the general belief, eating less meat and dairy products has a much larger impact on one's carbon footprint than eating locally.[11]

[8]Global warming of 1.5 °C, IPCC report, October 2018; on page 108 of the report, it is said that "if we want to have a 67% chance approximately to limit the global average temperature rise to below 1.5 °C, we had on January 1st 2018 about 420 GT of CO_2 still to emit."

[9]Thunberg warns EU ministers about world's declining carbon budget, https://www.euractiv.com/section/climateenvironment/news/thunberg-warns-eu-ministers-about-worlds-declining-carbon-budget/. Accessed 1 June 2020.

[10]https://ec.europa.eu/environment/eussd/smgp/ef_pilots.htm. Accessed 1 June 2020.

[11]Ritchie and Roser (2020).

A clear advantage of setting a PCB scheme is that it gives clarity to the huge challenge of diminishing the carbon footprint of Europeans.[12] Many NGOs and research organisations call for a carbon footprint per capita to be quickly reduced to 2 tonnes per capita. There are large differences—more than 100-fold—in per capita CO_2 emissions between countries. There are large inequalities in CO_2 emissions: the world's poorest contribute less than 1% of emissions, but will be the most vulnerable to climate change impacts.

Once a standard identifying methodology has been developed, a further major challenge faces the application of PCB. Technical and financial processes have to link the expenditure of individuals or households to their purchases and oversight processes developed. In addition, data safety and privacy issues will need to be taken into consideration. None of these challenges are beyond the scope of current technology but in practice they create daunting barriers to implementation. At best, in the foreseeable future, PCBs can provide a very useful set of indicators for enthusiastic carbon minimisers but cannot provide an effective regulatory mechanism for those choosing to exceed their allowance.

Stakeholders and the general public will never agree 100% on the do's and don'ts of climate action but there are ways to combine the interests of all to establish a grand coalition.

Governments, it is time for you to be creative!

A Collective Effort by Ambitious Governments

It took more than 20 years to start banning smoking in public places after evidence of the damaging effect of second-hand smoke on health had been confirmed by science. How many years should we wait for climate protection to become mainstream?

The climate demonstrations by young people rightly put the focus on the insufficiency of efforts by governments. Young people are perfectly right to ring the loudest of alarm bells. Young people are perfectly right to rebel against the "factory of slowness" that characterises the COPs and the way international climate governance operates. Young people are perfectly right to see the earth as a Zone To Defend, referring to the French neologism ZAD.[13] Young people are perfectly right to shake

[12]Ibid. 12.

[13]In French: Zone A Défendre, which evokes a sort of political squat.

existing governments. Young people are perfectly right to disobey peacefully—their future is threatened.

Governments face a unique challenge that has no equivalent in the history of humankind. Ursula von der Leyen was trying to evoke a can-do spirit when referring to Europe's man on the moon moment when presenting her EGD to the European Parliament in December 2019. Acting now or paying more later says the science, but the equation for engaged politicians is opposite, it is more costly now when thinking of the next election. The capacity of governments to be courageous, to develop a bold vision and to reform democratic processes to achieve it, is at stake, though in reality the challenge is much greater even than planetary colonisation!

States have an urgent and prime responsibility to set the right framework conditions to activate the necessary environmental transformation by different sectors of society. The challenge is that their longterm responsibility to future generations should not over-ride immediate democratic wishes. One way to reconcile this is for them to stimulate ongoing conditions for an interactive dialogue with civil society or, better, a co-creation of the climate solutions with all motivated actors. This initiative can only come from the top, government has to show the way.

The EU is encouraging direct citizen participation in policy processes. Two EU advisory bodies explicitly include the promotion of participatory democracy and civil dialogue in their remits: the European Economic and Social Committee (EESC) (1958) and the European Committee of the Regions (CoR) (1994). Practical ways to have a say include petitioning to the European Parliament, participating in public consultations or Citizens' Dialogues[14] to influence policy-design, and, last but not least, submitting proposals for new or changed EU legislation via the European Citizens' Initiative (ECI) that was introduced in 2011.[15] French President Emmanuel Macron initiated "European Citizens' Consultations" in 2018 to inform the discussions about the future of the EU at the European Council.

Is this enough? Have EU citizens the feeling that they really can be part of the overall picture? Not really. Despite these many, well-intentioned efforts, the distance between people and the final decision seems to have never been so great. The deliberative element is often missing. Recognizing that climate change requires a specific dynamic of information, communication and participation, the European Commission will be launching a Climate Pact in the autumn of 2020. Announced in its Communication on a European Green Deal,[16] it has three specific objectives: (1) To encourage information sharing, inspiration, and foster public understanding of the threat and the challenge of climate change, (2) To offer both real and virtual spaces for people to express their ideas and creativity and work together on ambitious action, both at individual and collective level; (3) To build capacity to facilitate grassroots initiatives on climate change and environmental protection.

---

[14]In 2015–20, the Europan Commission held almost 1600 dialogues with over 190,000 participants in 580 cities across Europe, see Citizens' Dialogues and citizens' consultations—key conclusions, 30 April 2019, European Commission.

[15]https://europa.eu/citizens-initiative/select-language?destination=/home. Accessed 3 June 2020.

[16]The European Green Deal; COM/2019/640 final.

Though the vast majority of people expect their government to take the lead role in articulating and financing the response to the climate challenge, this responsibility cannot be left to the mechanisms of government alone. The manifold dimensions of climate change implies that all people and organisations concerned by the production and consumption of goods and services must become active promoters of the needed transformation. Besides government and public authorities, it is therefore essential that the private sector and the general public are contributing. The climate responses must be both top-down and bottom-up; they must be the result of a whole-of-society transformational effort.

In its World Energy Outlook 2019 report,[17] the Executive Director of the International Energy Agency (IEA), Fatih Birol, insisted on bringing all political, economic and social forces together to meet the climate challenge: "The world urgently needs to put a laser-like focus on bringing down global emissions. This calls for a grand coalition encompassing governments, investors, companies and everyone else who is committed to tackling climate change".[18]

Can we look to the international response to the COVID-19 pandemic as an indicator of the appetite for global solidarity in the face of major challenge? There is certainly a lot to learn but the lessons are mainly about how not to behave and missed learning opportunities. Compared with international cooperation on preparing for and mitigating pandemics, the world of climate science and collaborative bodies is well advanced. 30 years of research and analysis have arrived at a broad consensus on what we need to do to avert the worst of the climate crisis and have agreed that comprehensive global action is vital. This preparatory framework was not in place to deal with the COVID-19 crisis. Exemplary actions were indeed taken by some states and these limited spread and eliminated infection at a national level. But without cooperative action at a global level, the good experience by some was rarely shared by the majority and the threat remained for all. Total public spending to counter the pandemic could have been much reduced with more and better coordination. The major obstacles to more cooperative action were a preference to develop and apply nationally created systems and a belief that the matter could be best resolved with a degree of independence from other countries.

At the present time, the various players tend to keep passing the ball back and forth to each other. Governments, public authorities, companies, the financial world, citizens and non-profit organisations: everyone is blaming the slow progress on the others' responsibilities and failures. Defining a common and collective approach with clear responsibilities is therefore an absolute must.

To give an idea of the scale of the respective contributions by the key stakeholders, it is useful to refer to what the signatory states to the Paris Agreement plan to realise in terms of reduction of their greenhouse gas emissions. This is encapsulated in the so-called Nationally Determined Contributions (NDCs), to use the United Nations jargon. The overall NDCs submitted by the countries are largely insufficient

---

[17]World Energy Outlook 2019, Flagship Report, The International Energy Agency, November 2019.

[18]The views of the IEA in their 2020 report in the light of COVID-19 are discussed in the next chapter.

to meet the 2 °C objective and certainly fall far short of the preferable 1.5 °C level. It is estimated that what the countries plan to do (a promise, not a commitment) only represent a third of the total endeavour needed. This would mean that, roughly speaking, both the private sector and civil society would have to deliver another two thirds.

A recent study by the French Carbon 4 company,[19] an independent consultancy specialising in low carbon strategy and climate change mitigation and adaptation, shows that people's efforts to engage society in the Transition, could represent up to one third of the Paris endeavour. In accordance with the Paris Agreement, a French citizen's average carbon footprint, which amounted to 10.8 tonnes of carbon in 2017, would have to decrease by around 80%, to a maximum amount of two tonnes of carbon a year by 2050. Carbon 4 has listed a dozen actions that individuals can take by their own initiative, not only small daily actions (using a reusable water bottle, using LED lamps etc.) but also more substantial behavioural changes (eating vegetarian and local food, giving up flying, systematic car sharing, purchasing three times less clothes, etc.). Yet put together, these efforts would be largely insufficient to meet the Paris objective as they would represent a mere reduction of 2.8 tonnes of carbon per person (out of a total of 10.8 tonnes). Interestingly, half of the reduction would be attained by shifting to a vegetarian diet, which shows the major importance of rendering food production and consumption more climate friendly.

What is clear is that there cannot be any substantial contribution by any single stakeholder group without the other groups acting similarly. Citizens will not be able to play their part on many of the actions listed by Carbone 4 if there is no upstream change of the boundary conditions set by governments. Citizens will not be able to consume more sustainably if the business sector does not redirect its production and marketing strategy, notably by reducing the offer of non-sustainable goods and services on the market and increasing those of sustainable ones.

For the time being, the people who consistently act for the climate are still a small minority. As alter consumers, they defend, sometimes very vigorously, a raft of ideas to profoundly adapt lifestyle. Past research in sociology, physics and other disciplines showed that small groups of people can succeed in flipping firmly established social conventions. When that happens, what was once acceptable can quickly become unacceptable, and vice versa. There is a controversy on the percentage of people needed to reach the critical mass or flipping point. Some argue that it is as low as 10% whereas others suggest it can be as high as 40%.[20] The figure can largely vary depending on the composition of the minority and their social claims. According to the American journalist and author, Malcolm Gladwell,[21] the tipping point is around 10%, giving strong hopes to those who think that climate consciousness will soon reach the tipping point. In France, the government has promoted a business initiative to engage more enterprises in the fight against climate change and reach the famous

---

[19]Doing your fair share for the climate?, November 2019, www.carbone4.com. Accessed 3 June 2020.

[20]Yong (2018).

[21]Gladwell (2001).

10% tipping point. Launched in May 2019 under the motto "10% for changing all",[22] the ambition is to federate all the actors of the Transition, as well as those of the social and solidarity economy. Though any political action aiming at reaching the tipping point is valid, one has to be careful in not confusing it with a threshold leading to real engagement to change ones' lifestyle.

What is central here is to realise a sort of cultural transition, which has to become collective to succeed. Changing the social norm in redefining the "cool" will be much easier if the call comes from key influencers in society, such as TV personalities, fashionistas or sport people, to list but a few. The ones who decide to apply the Transition for themselves, when the large majority of the population is still stuck in an obsolete model of development, must be given the social recognition they deserve as early adopters. A changing social norm can have a decisive impact on what people consider is their individual freedom. And for the change of the social norm to become quickly viral, it must convey the message that individual freedom is not diminished, it is just changed to better reflect collective responsibility.

Different stakeholders follow different ways and rhythms of action. Engaged citizens want immediate changes that governments are unable to deliver because of the length and complexity of decision making procedures. The Paris Agreement calls, however, for a quick delivery of actual emissions' reductions. It seems sensible to have a differentiated approach between those consensual measures that could be processed very quickly (e.g. pursuing the efforts to install more renewable energy or insulate homes and embed more participatory approaches in policy-making processes), and the measures posing sensitive acceptability issues, hence requiring more time for mutual concentration and even co-creation with civil society. Not proceeding this way exposes governments to the risk of pushing for measures that will quickly be revealed to be counterproductive through not being understood and endorsed.

Stakeholders and the general public will never agree 100% on the do's and don'ts of climate action but there are ways to combine the interests of all to establish a grand coalition. Governments, it is time for you to be creative!

What Governments Must Do

The role of governments is not only to propose efficient policy measures to combat climate change, but also to apply a modern governance approach that gives a clear direction and optimises the contribution of all civil society actors.

Painting a sombre picture of the way democracies care for the common good is an easy task. Whilst it is true that achieving an exemplary democratic process is still a work in progress recent years have seen such progress come to a halt or even be reversed. In the worst, but sadly not unique, cases democracies have been emptied of their meaning, certainly by the weakening of the idea of the common good, but also

[22]In French: "10% pour tout changer".

the erosion of checks and balances and the rule of law, the enslavement of public debate to collective passions and the tyranny of the short term which prohibits any constructive vision of the future. Taking the full measure of climate change through daring political decisions will not happen without reclaiming the essential ingredients of true democratic systems.

Governments' policies have so far been unable to mitigate climate change to the extent needed to preserve the long-term habitability of our planet. In part, this was due to the lack of financing and the lack of consistency between policies (Many governments support the take up of renewable energy whilst continuing to subsidise fossil fuels!). Besides rationalising policies, there is another vital obligation for governments: to strategically reinforce all the processes of governing climate change.

New governmental approaches are needed to stimulate all the processes of interaction and decision making among the actors involved. Governments hold the prime responsibility in the creation of a virtuous circle of positive actions for the climate. They are the initiators of the so-called top-down approaches and can, in this way, largely influence the way stakeholders and the general public do respond to the climate threat. All the institutional levels are to be involved, from the central government to the local authorities through the regional authorities. They all have the capacity to trigger action at their own level of competence.

If carried out holistically, the success of climate governance will be much more than the reduction of overall carbon emissions. To achieve this vital goal, governments will have to deploy an arsenal of policies and measures, offering new kinds of opportunities to create a more socially robust and economically balanced society.

Considering that there has been a wealth of studies and research on how best to reduce carbon emissions through innovative policies, the book principally concentrates on the way decisions are taken and civil society is getting involved.

People look naturally to their governments in time of crisis—will they be disappointed?

Climate Policies

Governments' tendency to opt for easy, technological or infrastructural solutions to repair climate change is, in reality, a sign that they lack political courage. Targeted taxation and specific measures mitigating destructive consumption are now needed.

For the Transition to succeed, different policy levers have to be used such as reinvesting in public transport, financing housing's insulation, or making sustainable food affordable. The number of proposals for new policies and measures that have been made by stakeholders (policy-makers, scientists, engineers, corporations etc.) is huge. However, their level of implementation is still largely insufficient, scattered and uncoordinated.

Even applying the relatively consensual levers seems exceptionally difficult. We should reduce fossil fuel consumption—but coal consumption has increased by 23%

in the last 10 years at global level. We should stop deforestation—this has increased by 50% in the last 10 years. There are also other approaches that are more controversial such as a reduction in air transportation, yet this has increased by 64% in the last 10 years to 2019; cutting back on meat consumption (+11% in the last 10 years) or the application of economic, education and contraception programmes to mitigate population growth (+200,000 persons every day).

Stopping deforestation faces many hurdles among which is the desire of forest owners to maximise their profit by transforming a rich ecosystem into poor arable land. The abandonment of fossil fuel subsidies, responsible for around 55 billion euros of public money every year in the EU,[23] is a priori another consensual measure. Yet, not a single EU member state has so far spelled out a comprehensive plan to phase out fossil fuel subsidies.[24] EU countries and Norway earned more than €400 billion in revenue from oil and gas taxation in 2017,[25] underlining the difficulty for governments to end the subsidies despite a commitment taken at the G20 ten years ago to eliminate them.[26] The question is becoming even more politically sensitive if palm oil production and use are brought into the picture. At the global level, eliminating these subsidies would raise government revenue by 3.6% of global GDP and help to cut GHG emissions by more than 20%.[27]

Into this situation has come what might be termed a discontinuity of unknown impact in the form of the global recession resulting from the COVID-19 pandemic. Oil prices collapsed, dragging down the cost of both coal and gas by 30–50%. As global transport and many forms of industrial production slowed or ceased, demand for fossil fuels dropped to exceptionally low levels. Fatih Birol, the executive director of the International Energy Agency said: "The plunge in demand for nearly all major fuels is staggering, especially for coal, oil and gas. Only renewables are holding up during the previously unheard of slump in electricity use." The IEA predicted a drop of 3bn tonnes in CO_2 emissions in 2020, a reduction of 8% in what it termed "the biggest shock to the global energy system in more than seven decades"[28] Even the exceptionally well informed IEA concluded in its April 2020 Global Energy Review that, "The energy sector that emerges from the Covid-19 crisis may look significantly different from what came before."[29]

Even harder is to apply more controversial measures such as reducing population growth, which is taboo in a majority of countries. In the period 1979–2015, China decided to control its population growth via drastic measures, going well beyond

---

[23]Study on energy prices, costs and subsidies and their impact on industry and households, Cambridge Econometrics, Directorate-General for Energy (European Commission), Enerdata, Ludwig Bölkow SystemTechnik (LBST) and Trinomics B.V; research contracted by the European Commission, 14 January 2019. The research covers the years 2014–16. The level of 55% stayed broadly stable over that period.

[24]van der Burg L et al. (2019).

[25]According to the International Association of Oil and Gas Producers.

[26]G20 Summit in Pittsburgh, 24 September 2009.

[27]New sustainable economic model, opinion by the EESC, OJ C 81, 2.3.2018, pp. 57–64.

[28]https://www.iea.org/reports/global-energy-review-2020. Accessed 3 June 2020.

[29]Ibid. p. 42.

educational programmes. In 2006, Zhang Weiqing, Director of the National Population and Family Planning Commission, admitted that the so-called one child rule has enabled China to avoid approximately 400,000 births. However, although the policy had some effect on reducing consumption and growth it is blamed for having long-term devastating consequences socially with one child taking care of 2 parents and 4 grand-parents and 33 million more men than women in 2014.

Similarly, the calls for limiting our meat consumption are probably just as provocative as ending smoking was 60 years ago. Meat's role in the diet is deeply cultural in many countries and eating more meat is seen as a sign of an improving lifestyle so the resistance to changes of that nature is not to be underestimated.

The range of measures that tend to be preferred by decision-makers so far are, in the main, related to technology development. The deployment of renewable energy sources to replace fossil fuel is an excellent example of a successful recourse to technological development to reduce the carbon footprint of energy production. There is, however, a tendency to exaggerate the actual role that could be played by various technologies in the mitigation of greenhouse gas emissions. Our society's belief in progress is fuelled by the continuing belief in uninterrupted and promising technological innovation. Part of this overconfidence is caused by the success of the technological approach which was used for stopping the thinning of the ozone layer. The protective measures for the ozone layer that were taken under the Montreal Protocol indeed implied a significant amount of technology development. As stressed by the American journalist Richard Heinberg,[30] technology solutions tend to be favoured because they are less complex to deploy than other kinds of measures requiring more in-depth understanding of their societal impact. We all like technologies as they have demonstrated their continuing capacity to cure diseases and expand food production and there is no reason that it should not be like that into the future!

Certainly, renewable energy must continue to be implemented in residential areas, factories and commercial centres. Likewise, energy efficiency measures will continue to require more sponsoring and promotion by governments. But to resolve a much more global and complex issue than the thinning of the ozone layer, one cannot count on technology development alone. Governments must also refrain from saying that they will wait for the next, better technology of tomorrow before making the needed investments. A plethora of international studies have said very clearly that climate change can be resolved with the available clean technologies. In its 2050 climate neutral strategy, adopted in November 2018, the then EU Commissioner for Climate Action and Energy, Miguel Arias Cañete said: "We propose a strategy for Europe to become the world's first major economy to go climate neutral by 2050. Going climate neutral is necessary, possible and in Europe's interest. It is possible with current technologies and those close to deployment".[31]

[30] Heinberg (2017).

[31] A Clean Planet for all—A European strategic long-term vision for a prosperous, modern, competitive and climate neutral economy; Communication from the European Commission; COM (2018) 773; 2018.

Besides technological approaches, there is a raft of additional policies and measures that are to be considered for designing future climate plans. In the first place, there is a huge potential to be tapped from nature-based carbon offsets like forestry and land use activities. Such solutions must be increasingly valued and leveraged by the policy-makers as they are very cost effective. Besides these nature-based solutions, there are essential "horizontal" measures such as education and vocational training, research and development, financing, sharing of good practices, or international cooperation that have also to be triggered.

Most essential is social innovation which deserves much greater policy, regulatory and institutional support. Social innovation embraces new social practices that aim to extend and strengthen the role and capacity of civil society. The social processes of innovation, such as open source methods and techniques, fit perfectly well into the collaborative spirit of developing climate action in common. Many innovations which have a social purpose like activism or online volunteering are also nurturing and diversifying the modes of practical action. Social innovation can be considered as the means through which the Transition could become a societal Transition that encompasses and engages an entire society. It has a great potential to lead to significant behaviour and lifestyle changes, which are at the core of successful climate mitigation policies. For all these non-technological measures, there is a need for clear, efficient and effective underlying legislation that can provide for immediate implementation.

The regulatory barriers to any kind of positive climate action have to be removed. So, for example, the possibility to connect vehicle batteries to the grid in order to get access or provide energy from/to the network has to be affirmed—an operation that is still forbidden in several European countries. One of the key contributions to climate protection being the change of consumption patterns, existing legislation has to be amended to provide more incentives for adopting sustainable lifestyles. There is a pressing need to restructure the legislation to promote ways of living that take a better account of tomorrow's generations, as proposed in the ten action point manifesto.

Among the non-technological approaches, adjusting the pricing of goods and services on the basis of their economic, social and environmental impact is seen by many as a must. A new taxation system will have to be imagined for the twenty-first century. Energy and primary resources, and their per-capita use, have to be taxed to the extent of their impact on carbon emissions, using either a carbon emission trading approach (such as the EU Emission Trading System[32]), a carbon tax or a mixture of the two. In the same way, the taxation on consumption has to be increased while the taxation on labour has to be reduced, as recommended by the OECD.[33] The VAT regime will equally have to be fine-tuned to the inherent carbon emissions of goods

[32]The EU Emission Trading System (ETS) has its price of carbon evolving over time as a function of the reducing cap and the market-based evolution of demand for carbon allowances, see https://ec.europa.eu/clima/policies/ets_en. Accessed 3 June 2020.

[33]A Comparison of the Tax Burden on Labor in the OECD, Elke Asen, FISCAL FACT No. 655 May 2019.

and services, notably by stimulating the reuse of recycled material or the reduction or exclusion of plastic components.

On top of developing ambitious climate mitigation policies, governments also have the duty to take adaptation on board by ensuring that new developments do reinforce societal capacity to withstand climate events of different kinds. The economic costs of adaptation to climate change are likely to cost tens of billions of euro annually for the next several decades, though the exact amount of money needed is unknown and dependent on the effectiveness of mitigation. What is certain is that these costs will continue to rise steeply in the future if mitigation measures are inefficient or insufficient. With its higher ratio of coast to landmass than any other continent or subcontinent,[34] Europe will have an increasing number of properties that will no longer be insurable because of flooding risks and numerous communities—large and small—will be threatened.

Taxation and pricing are key in reorienting goods and services towards sustainability but to be widely accepted the reform of taxation will also have to be designed in such a way as to dissuade the wealthy from continuing to over-consume. This is a not insignificant challenge!

The Barriers to Effective Climate Governance

European citizens are tired of government reforms. The word reform should normally mean an improvement in everyday life but, in a Europe losing ground against international competition, most recent reforms have often had a negative effect on well-being.

New and good policies are not enough to deliver the expected impacts. Governments have to adapt their governance and top-down approaches to provide stronger and quicker responses to the huge challenge of climate change. Good governance principles include strategic vision, transparency, equity, accountability, participation and responsiveness, effectiveness and efficiency, as well as respecting the rule of law. All these principles are equally important for achieving good governance of climate policies. Engaging the general public in the elaboration and implementation of these policies and guaranteeing equity through genuine burden-sharing by those most able to afford it, can be considered as key determining factors.

Are democratic processes fit to handle the climate issue? This question is on the lips of many observers, in particular when looking retrospectively at the sequence of previous attempts and their poor end result. Some of the reasons why democratic systems may be less apt to respond to the climate change challenge are:

- an excessive focus on the short term, i.e. a form of short-sightedness reflecting electoral cycles—the dominance of political parties (also named particracy), versus other forms of representing people's interest

[34]Cuper (2014).

- the conservatism of the large majority of political parties and their parochialism (the inherent trend to be narrow in scope),
- the genuine conflict facing voters when choosing who best can deliver their priorities—the subtle capture of political decision-making by the financial and economic elites.

In its current stage of evolution, democracy does not seem adequate to deal with complex global problems. Yascha Mounk, political scientist at Harvard university, argues[35] that it is a big mistake to believe that democracies will last forever. That could be a similar category of error committed by Francis Fukuyama when he judged that liberal democracies will naturally impose themselves after the fall of the Berlin wall in 1989.[36] The reality is that many democracies are facing hard times. Legitimacy losses, failures to guarantee equality among citizens and resolve big transnational issues, raging populism, disobedience or even violence becoming commonplace are all the symptoms of suffering democracies. Populism is the most emblematic expression of the erosion of democratic systems, leading to harsh antagonism and polarisation. Populism is a major source of concern when it comes to addressing climate change which primarily requires collective solutions.

In EU countries, support for democracy remains very strong but there are also clear signs of a demand for change, especially from younger generations. 68% of respondents agreed that "there is no substitute for the democratic system, it is the best possible system".[37] Younger respondents are less supportive though: 62% of those aged under 35 agree compared with 76% of t those aged over 60. Amongst those saying that democracy is the best system, a strong minority (41%) think that democracy is not working well in their own country. And among the renewal options, there is strong support for representative democracy (87%) and for direct democracy (66%).

The hyper-cautious and short-termism approach followed by democratic governments seems to be widespread. In an article entitled "Is democracy the problem?"[38] Thomas Carothers said: "Successive U.S. Administrations have proven woefully unable to focus sustained attention on a raft of major longterm challenges—whether it is infrastructure decay, the role of entitlement spending in the U.S. budget, or climate change—and unwilling to craft reforms that inflict short-term pain for the sake of long-term gain. The disproportionate influence of wealthy individuals and corporations in the U.S. legislative process is a well-known reality. With respect to political competition producing divisions and conflict, the U.S. political system is indeed beset by a high degree of polarisation and a correspondingly low sense of common purpose". The response of the Federal Government to the COVID-19 crisis illustrated many of these points.

---

[35]Mounk (2018).

[36]Fukuyama (2006).

[37]New Global Survey—Democracies under pressure—volumes 1, 2 & 3, FONDAPOL, May 2019.

[38]Published Carnegie Endowment for International Peace https://carnegieendowment.org/. Accessed 16 January 2019.

Political parties tend to court voters by defending conservative programmes that do not hurt people's current situation. The interests at risk in relation to climate change are huge and it is therefore no surprise that the political decision-making process has been and will continue to be largely influenced by financial and economic elites. The latter have so many vested interests in current assets that they actively dissuade governments from taking any action that would jeopardise them.

Climate change has recently entered into the programme of almost all political parties, with the noticeable exception of the extreme right and of many nationalist parties. As a whole, green parties remain the only ones asking for climate to be treated centrally alongside other pressing environmental issues though some parties on the centre-left (social democrats) are progressively integrating climate change in the core set of their proposals. The key influence of political parties in setting the political agenda of governments is another reason why climate issues have difficulties in penetrating the core of governments' programmes. The problem of the dominance of political parties or particracy has been stressed by many authors, including Wilfried Dewachter who has shown that in Belgium real power does not coincide with constitutional power[39] In Belgium, a majority of the most powerful political positions are not foreseen by the Constitution. This is particularly the case with the political party leaders who have considerable freedom to conduct the party's affairs without being bound to parliamentary debates or pressures from civil society.

Part of the reason why political parties do not dare to come up with courageous measures is the mistaken belief that such measures will hurt the voters or require unacceptable change. In other words, politicians tend to associate climate measures with measures that inevitably reduce the quality of life of voters or require them to make changes for which they are unprepared. In their eyes, proposing such measures in parties' programmes would mean losing the elections. Such an attitude is short-sighted and prevents comprehensive, top-level consideration of green policies. Making politicians aware that effective and efficient pro-climate measures can also benefit the entirety of society and the economy must be a first step in changing practices. It could be the case that these attitudes will be moderated in the light of the COVID-19 crisis. For this to happen, a link will need to be made between the short-term, highly interventionist approach to the public health crisis and the longer-term but equally impactful climate emergency. Furthermore, combining the Transition with social justice is not only possible it is required if we do not want concerns about equity to further delay action.

Democratic systems are normally constructed in such a way that general enthusiasm for all popular viewpoints can be tested through elections. The reality is that climate protection is far from being the only aspiration of voters. Politicians are confronted with a cacophony of opinions and desires. Though the people's concerns about climate change are on the rise, they are far from dominating. It is therefore very difficult for politicians to hear amidst the competing noise the most essential

---

[39] De mythe van de parlementaire democratie. Een Belgische analyse, ed. Acco Uitgeverij België (1 February 2001) (Dutch).

demands, the ones that require priority actions. Two challenges are faced by politicians. Firstly, how to convince the electorate that the short-medium term transition period involving radical change is both necessary and essential and secondly to demonstrate that globally equivalent action is also going to happen.

Even in cases where governments are ready to act to protect the climate and society, it may happen that the tabled proposals are deemed not acceptable by the people. Refusal can have different causes such as the non-involvement of civil society in the design of the proposal or the excessive imbalance of benefits/inconveniences across the population. Changing people's mind regarding the necessity, validity and medium-term benefit of climate reforms is another important challenge.

To recreate a positive connection between governments and the general public is at the heart of climate governance.

What People Must Do

Should the role of citizens be merely to call for government's action without themselves contributing to making these actions efficient and effective?

Besides forcing governments to take their statutory responsibilities, people have to become aware that a large part of the solution has to come from them also. Governments' plans should normally imply a more active and widespread participation of citizens in the realisation of national, regional and local objectives. Measures to make democracy more participatory or to make climate measures more acceptable or even fun, will not deliver their objective if people do not respond enthusiastically to them. The difficulty will be to have all the strands of society taking a conspicuous role in the collective endeavour. Furthermore, the diverse policies and measures decided by governments will not deliver their full potential if all people are not aligning their ways of behaving and consuming with the core objectives of these policies and measures.

At the source of people's engagement is their belief that it is worth a try. Movements such as collapsologists or survivalists would tend to consider that only preparing for the end of society as we know it is worth doing. Survivalism, which tended in the past to be essentially fed by religious and apocalyptic types of theories, has now entered a more rational and even scientific era[40] with climate change foremost as the supreme threat. The message of scientists, especially of the IPCC, is certainly pessimistic but it cannot be confused with saying that it will bring about societal collapse tomorrow and that any effort to mitigate risks has to be abandoned. The same confusion exists with the idea of progress. Progress has been the main driver of societal development in past decades but seems to have lost its capacity to make people believe in a better tomorrow. Actually it is another idea of progress that needs to be socially constructed and this has to involve defining more clearly what we want from progress. The very first step in having people (re)doing positive

[40] Vidal (2018).

actions for the climate would be to recreate hope and confidence in the future. A possible starting point is a meaningful and viable vision of prosperity for people and for the planet. An EESC opinion[41] amplified the proposition that, "prosperity is not simply a matter of production or income. It is more precisely defined in terms of the possibilities that people have to flourish and to achieve their full potential."[42] Here the role of politicians is again essential, as they are the organisers of the democratic space. Opening doors and windows, and inviting citizens to a dialogue on how the future can remain inclusive, secure and desirable, is the powerful first step that has to be undertaken by politicians, the number one gesture before people come to feel that they, themselves, have to get involved.

It is also important to affirm that a lot can be done by people, not a meagre contribution but a substantial one. In his essay entitled "We are the weather: saving the planet begins at breakfast",[43] the US writer Jonathan Safran Foer stresses the citizens' responsibility towards climate change and shows how practically everyone can contribute to reducing carbon emissions by modifying their diet. Saving the planet, he admits, will require that each of us deeply reconsiders what we do and do not, daring to challenge the deep psychological barriers that maintain us in a comfortable status quo. And the abovementioned Carbon 4 study singles out that changing diet is a very efficient way to drastically reduce one's carbon footprint. It is to be hoped that the response by the public to the lockdown measures proposed across many of the world's nations in response to the COVID-19 pandemic will indicate that personal responsibility in observing preventative actions can make a significant difference.

Those acting on their own initiative are still a minority today. For the largest majority of European citizens, "the end of the month is a higher preoccupation than the end of the world". Transforming the action of some into a collective endeavour, bringing together the entire population to rally for the agenda for change, will require ingenuity and perseverance. People's motives for action and adapting to societal challenges are very diverse. From the ones often called "social justice warriors"—highly committed activists who demand similar action from others, to the truly sincere but confused environmentally concerned citizen, there is a large spectrum of reasons why people suddenly decide to "do something".

Fighting climate change is not in contradiction with exerting one's individual freedom. Often, what people initially see as a major constraint, becomes rapidly an automatic and well-accepted gesture. Fastening the seat belt when driving is the perfect illustration of a legislative measure that protects and has gained quick acceptance.

There is no reason to believe that every citizen cannot become an aficionado of climate protection!

---

[41]The sustainable economy we need (own-initiative opinion), rapporteur: SCHMIDT P, www.eesc.europa.eu/en/our-work/opinionsinformation-reports/opinions/sustainable-economy-we-need-own-initiative-opinion. Accessed 3 June 2020.

[42]Jackson (2009), Eathscan also Raworth (2017).

[43]Foer (2019).

Consumption Is the Hardest Nut to Crack

Questioning unsustainable consumption is an obvious starting point. We all feel, from time-to time, that we buy too much 'stuff'. But it is probably the most difficult feature of our lives to tackle!

For governments to stand a chance of their climate efforts succeeding they must appeal to people as citizens rather than as consumers. Emphasising a common duty to make society flourish will involve a new emphasis on rights and responsibilities. The consumer needs to be transformed into the contributing citizen, enabling society to draw on the well of resources that is contained in all of us.

Questioning consumption not only threatens actions by which people affirm their identity but is tantamount to challenging a well-established system with many vested interests. There are two high hurdles to overcome. Firstly, all governments favour consumption as consumption correlates with the growth of GDP. Secondly, both people and governments favour consumption as consumption is perceived to correlate to happiness. In today's consumer society, the mantra is 'I shop therefore I am' and if I feel depressed then the answer could be retail therapy.

There are however many misconceptions about the linkage between consumption and happiness. The political scientist Stefano Bartolini[44] has formulated interesting propositions which may reduce consumption by increasing collective happiness.[45] People, he says, are overconsuming because they are unhappy and because of the "social capital" crisis. Consumerism is regarded as a mechanism of compensation for the poverty of human relations and for the feeling of loneliness. Public authorities have therefore a great role to play in reconstructing social capital, essentially via measures relating to urban planning and education. Redesigning cities to create more natural links (for example car-free) to public spaces where people can meet, is one possible way. Profoundly changing the education method of schools, from a competitive place of abstract learning to a collective place of social learning, is another way. Informing young people that dialogue with others is one of the most fulfilling experiences in life, well ahead of owning and acquiring goods, is key.

The works of Stefano Bartolini are in stark contrast to the often-used assertion that it is the race for happiness that force people to overconsume. Reversing this well-established logic will demand educational effort and it is not taken for granted that, in today's individualistic society, people will adhere to the idea that by creating more links with others they will reduce their basic impulse to consume.

A large majority of people indeed believe that they should have more income to sustain more consumption and thereby reach the fullness of well-being. The American economist Richard Easterlin demonstrated in the 1970s[46] that the increase in happiness when income is increasing is not long-lasting. It functions when a low income is being increased initially but further increases tend to have little or no

---

[44]Chief editor of the Italian Political Science Journal and of the Italian Political Science Review as well as Professor at the European University Institute of Florence.

[45]Bartolini and Sarracino (2014).

[46]Rojas (2019).

impact on the felt well-being. So gaining more money to buy more goods, cannot be confused with a permanent happiness recipe! But consumption has many other drivers. It is also the case that "consumer goods provide a symbolic language in which we communicate with each other, not just about raw stuff, but about what really matters to us: family, friendship, sense of belonging, community, identity, social status, meaning and purpose in life."[47]

For this and other reasons consumption is on the rise, not only in developing countries but still in developed countries. The very fact that the latter have already benefited from decades of intense material consumption, exemplifies the apparent drive to acquire and consume over long periods of time. In its 2015 State and Outlook report, the European Environmental Agency (EEA) stressed that "Household consumption expenditure in Europe increased by 23% in 1996–2012 in Europe, contributing to increases in some environmental pressures. Reducing the impacts of European consumption requires fundamental changes in lifestyle, including in the size and location of dwellings, transport systems and diets".

In a nutshell, one could say that both poor and rich people aspire to consume more tomorrow than they do today. The reasons for this seemingly endless consumption differ: whereas the poor initially seek to satisfy needs before aspiring to the lifestyles of those who are richer the affluent see in high consumption the affirmation of their social status.

The caricature of hyper-consumerism is so-called Black Friday with its heavy marketing and its hysterical behaviour. In France, the 2018 Black Friday meant a 71% increase in sales compared to a normal Friday. The shoppers are principally women and clothes constitute the majority of purchases. Seeing garments as disposable products is problematic as the clothing sector itself is responsible for no less than 10% of total carbon emissions. The total GHG emissions from textile production stood at 1.2 billion tons annually in 2017 and this was more than those of all international flights and maritime shipping combined.[48]

Typical of the way consumers trap themselves is their general awareness that they are buying more than they need but they cannot resist acting in this way. Wardrobes of European households are full of clothes that were worn only once. On the very day of the 2019 Black Friday, the fourth world climate strike (re-named Block Friday or Green Friday) was held in 2300 cities in 152 countries,[49] bringing onto the streets thousands of climate activists while shoppers and e-shoppers spent a significant part of their day in getting the best deals.

There are, however, a certain number of positive signals that indicate the growing desire to consume differently. With the increased understanding of environmentally sensitive models of society, people grasp more precisely what it means to consume in a more sustainable way. The survey by the French Observatory of Society and Consumption highlights that 56% of people claim that they are willing to consume

[47] Jackson (2009).

[48] The price of fast fashion, Nature Climate Change, 8, 1, 2 January 2018.

[49] According to Reuters.

"better", with 36% ready to consume less but better. Only 21% declare that they want to consume more.

Comparing poor and rich people's consumption patterns leads to questioning whether it would be feasible to distinguish needs on the basis of an agreed set of criteria. To eat or to protect oneself from the cold are essential needs (but unmet for millions of people who desperately lack access to food and warmth). To travel for purposes that go beyond access to food, employment, healthcare and other public services, is not an essential need but can be considered as desirable. Someone who has not travelled widely for leisure is regarded as lacking an important dimension of modern life but it cannot be said that this person has missed something essential in sustaining life. In referring to these desirable goods and service, André Gorz defined them as "qualitative needs".[50] Leisure travel is a need that is culturally constructed and that has become important over time (but not vital!). And the reason why carbon intensive traveling, as a perceived essential need, has become a focal point for experts and media is related to its disastrous environmental impact.

In the context of reducing the footprint of consumption, what can be considered as an essential and therefore legitimate need is likely to be a highly disputed item of debate. Climate change and the necessary Transition are forcing us to re-evaluate the way essential needs are met. Can there be needs that should not be satisfied because they are too selfish and imply unsustainable production methods? That is the very controversial proposal made by Professor Razmig Keucheyan in his recent book on the ways to move away from consumerism.[51]

In today's tense debates on the balance between social and environmental justice, symbolic but minor issues can suddenly become headline news. In the EU, plastic straws will soon be prohibited but travel keeps rising. What rich people can afford in terms of green consumption, such as buying more expensive electric vehicles, does not bring a significant contribution to alleviating climate change and remains largely out of reach for the majority of the population. The resolution of the climate challenge cannot merely be achieved by the "ecology of the rich", as stated by Peter Dauvergne in his book on "Environmentalism of the Rich".[52] Aude Vidal speaks of "egology" when referring to what people with economic, social or cultural capital can afford in comparison with the rest of the society.[53]

Consumer societies around the globe all depend on a basic premise—a continual process of purchase, use and replace. This is encouraged by social drivers which, after a certain level of basic need has been met, promote self-realisation through consumption. In this sense, Marx was right[54]! He understood that (market) capitalism, now the dominant economic form even when state-sponsored, is based on the frequent consumption of short-life, disposable or novel goods. By generating artificial 'needs', it also creates forms of alienation caused by the essential needs becoming confused

---

[50] Bowring (2000).

[51] Keucheyan (2019).

[52] Dauvergne (2016).

[53] Vidal (2017).

[54] Ökonomisch-philosophische Manuskripte aus dem Jahre 1844 or Pariser Manuskripte.

with artificially created ones. Marx also identified the negative impacts of capitalism on nature through pollution and the exhaustion of resources.

It is not easy to opt out of the prevalent socio-economic patterns of consumption. Even replacing more frequent purchases of low cost goods with less frequent purchases of higher quality, more durable products has its limits. The reality is that not only do we buy too much 'stuff' we can also be dependent on everyone continuing this process to maintain our own livelihoods, as the covid crisis exemplified. Mahatma Gandhi may have said 'Live simply so that others may simply live' and it certainly became a popular way of highlighting overconsumption by Western countries in the 1970s. But fifty years on it has acquired a different meaning. Poverty remains a stain on our humanity but billions have left its grip in the last two generations thanks to massive production in developing countries and massive consumption in developed countries. Now an additional challenge is not to impoverish future generations by our overconsumption today.

Looking at what people concretely do for the climate, an important gap continues to exist between developing an environmental consciousness and taking positive actions to support the Transition. The preferred actions are the ones that have little financial implication or impact on lifestyle and are regarded, for good and bad reasons, to be emblematic of one's changed consumption behaviour. Today the non-plastic, reusable water bottle best symbolises the quick emergence of an environmental consciousness in the European population. The way it has diffused so quickly is certainly a matter of satisfaction. Organisations like the European Economic and Social Committee (EESC) have offered such bottles to their Members and staff and put water fountains at their disposal, after having banned plastic bottles in meeting places. However, as revealed by the Carbon 4 study on people's contribution to climate protection,[55] these small daily gestures will be quite insufficient to reverse current trends. Changing diet, giving up air travel and minimising clothing purchases are the kinds of decisions that have, by contrast, very significant impacts in the medium term, but they require a level of commitment far greater than displaying a superficial, socially varnished, environmental consciousness.

In a report on the assessment of the environmental impact of consumption,[56] the International Resource Panel established under the auspices of the UN Environmental Programme (UNEP) has concluded that the production of goods and services for households is the most important cause of GHG emissions, with food, housing and transport dominating the environmental impacts. With household consumption determining 60% or more of the OECD countries' GDP,[57] governments are warned that any ambitious climate governance will have to rely heavily on curbing and reorienting consumption.

[55] Dugast and Soyeux (2019).

[56] Assessing the environmental impacts of consumption and production—priority products and materials, UNEP, International Panel for Sustainable Resource Management, Working Group on the Environmental Impacts of Products and Materials 2010.

[57] https://data.oecd.org/hha/household-spending.htm. Accessed 1 June 020.

New ways, therefore, need to be found to make it easy for people to contribute more spontaneously to climate protection. Internalising the cost of carbon emissions and reflecting them fully in the price is a promising route that could take inspiration from similar practices relating to the recycling of products. In the EU, the purchase of different categories of products includes the payment of a recycling fee. It has the advantage of being relatively painless as the fee is often an order of magnitude cheaper than the total price of the product. For example, the EU legislation promoting the collection and recycling of electrical and electronic equipment, which has been in force since February 2003, provides for the creation of collection schemes where consumers return their used waste equipment free of charge. The objective of these schemes is to increase the recycling and/or re-use of such products. These EU schemes have been revealed to be quite effective and similar approaches could be used to internalise the offset costs for carbon emissions in the price of products and services. Here again, the first step is to be taken by governments by imposing a carbon offsetting fee to a category of products and services that are proved to be carbon intensive.

According to IATA (International Air Transport Association), just 1% of passengers offset their carbon emissions when flying. Over 30 IATA member airlines have introduced an offset programme either integrated into their web-sales engines or to a third party offset provider. The 140-fold growth of carbon offsetting between 2008 and 2018, with 430 million tonnes of emission reductions generated since 2005 sounds good until it is set against the large, continuously rising carbon emissions due to air transport. In 2018 alone aviation generated 918 million tonnes! Having the carbon fee automatically included in the price of an air ticket, would be a true improvement. A very positive signal has come from easyJet in November 2019 with their decision to become the first major airline to offset the carbon emissions from the fuels used in flights. The impact of COVID-19 on air transport was such that the likely recovery period for that sector could well extend for several years. Unfortunately the majority of public subsidies decided by governments from April 2020 was not aligned with mandatory climate-friendly conditions of budgetary support. However, one positive exception was the case of Air France/KLM that was substantially helped—around €10 billion in loans and guarantees—by the French and Dutch governments with some modest carbon-reducing conditions attached.[58]

Each day, there is a new business or concept popping up with the word sustainability attached to it and established businesses which chose to emphasise that angle. This combination of environmental awareness and social action is not new and can be traced back to the start of the co-operative movement in the 1840s. The organic food movement was founded in Germany and the UK nearly a century ago and from the 1970s, the Fair Trade movement has spread into most developed and many developing countries. All these movements are increasingly recognised as facilitating and growing a greater awareness of social and environmental issues at consumer's level. Whilst they all have national frameworks and representative bodies at a global level,

---

[58] https://www.euractiv.com/section/aviation/news/air-france-klm-gets-virus-bailout-green-conditions-still-pending/. Accessed 1 June 2020.

they also enable local action by individuals or groups. Alongside these movements, hundreds of certification labels have sprung up offering various assurances about individual products or services. The consumer is now well served, and probably confused, about the numerous guarantee marks related to sustainability which can be attached to everything they buy, from holidays to clothing, wine to chocolate, energy to packaging.

Encouraging people to adopt sustainable consumption necessitates making it easy, local and rewarding.

References

Bartolini S and Sarracino F, 2014 It's not the economy, stupid! How social capital and GDP relate to happiness over time.

Bowring F. (2000) Ecological Crisis and the Limits to Economic Rationality. In: André Gorz and the Sartrean Legacy. Palgrave Macmillan, London.

Cuper, S. "Why Europe works". Financial Times 24 May 2014.

Dauvergne P, (2016) Environmentalism of the rich, MIT Press, Cambridge, Mass.

Dugast C and Soyeux A, Doing your fair share for the climate?, Carbone 4, November 2019, www.carbone4.com.

Foer J, (2019) Hamish Hamilton, London.

Fukuyama F (2006 ed.) The End of History and the Last Man, Free Press, New York.

Gladwell M (2001) The Tipping Point, Little, Brown and Company, New York.

Heinberg R, There's no app for that (2017) Post Carbon Institute, Santa Rosa.

Jackson, T. (2009). Prosperity without growth. Economics for a finite planet. Earthscan,.London.

Keucheyan R (2019) Les besoins artificiels. La Découverte, Paris.

Lenzen M et al., Nature Climate Change 8, pages 522–528 (2018).

Mounk Y, (2018) The People vs. Democracy, Harvard University Press, Cambridge, Mass..

Nachmany M and Mangan E (October 2018) Policy brief "Aligning national and international climate targets", Grantham Research Institute on Climate Change and the Environment.

Raworth K, (2017) Doughnut Economics, Cornerstone Digital.

Ritchie H and Roser M (2020)—"Environmental impacts of food production". Published online at OurWorldInData.org. Retrieved from: 'https://ourworldindata.org/environmental-impacts-of-food' [Online Resource] Accessed 1 June 2020.

Rojas M, (ed) (2019) The Economics of Happiness. Springer International, Cham.

van der Burg L et al. (2019) Fossil fuel subsidies in draft EU national energy and climate plans; Friends of the Earth Netherlands and Climate Action Network Europe and Overseas Development Institute.

Vidal A (2017) Egologie: écologie, individualisme et course au bonheur, Le monde à l'envers, Paris.

Vidal B (2018) Survivalisme, Arkhé.

Yong E, The Tipping Point When Minority Views Take Over, The Atlantic, 7 June 2018.

Chapter 4
Ten Measures to Make Climate Governance Fit for Purpose

There are many ways to adapt democratic governance to the pressing challenge of climate change. This book concentrates on the ten most practical governance changes that could deliver a quick and substantial response to climate change. Some of these changes have been requested by youth movements defending the climate cause, others have been long-standing demands from a wide range of civil society organisations, many have been adjustments advocated by environmental scientists and economists during the last half-century. Some of these action points have appeared in the manifestos of political parties around the world—and not just those with a predominantly Green objective. The relevance and consensus on these ten changes have been dramatically enhanced by the covid crisis and its aspiration towards a "new world".

There are innumerable concepts in the toolbox labelled 'Climate Action'. Policymakers and decision takers are therefore never short of good ideas from which to select, the challenge is always to identify those which will receive the support of the general public, stand a high chance of being effectively implemented and are designed in such a way as to benefit all sections of society. For this high societal return to become reality, governance of the decision-making and decision-implementing process has to receive the same level of care as the policy and project design. The first eight action points in the Transition manifesto specifically focus on how governance can be adapted in all its inherent dimensions: leadership, deliberative and inclusive approaches, involvement of people, youth, civil society and industry, wide communication to support people's and organisations' decision-making, open-mindness to consider any politically sensitive measures, combination of technology and nontechnology approaches, long termism, holistic dimension, etc. The last two changes concern specific strategies to support a reform of economic policies and promote more sustainable consumption; though not being governance changes per se, they are so far-reaching and politically demanding that their inclusion in the list of the ten was deemed justified.

E. Ponthieu, *The Climate Crisis, Democracy and Governance*,
SpringerBriefs in Climate Studies,
https://doi.org/10.1007/978-3-030-58127-5_4

Ambitious States in the Driving Seat

The reconnection of political decisions with the preservation of the long-term common good is at the heart of this manifesto. This requires governments and politicians to display strong and ambitious leadership reflecting a comprehensive package of measures which will tap into the latent resources of society. The core aspects of governance needed to enable progress across the 2030 SD Agenda and meet the 17 SDGs—inclusive, participatory and representative decision-making; effective, accountable and transparent institutions; responsive, access to information and protection of fundamental freedoms—are similarly required to achieve SDG 13 (Take urgent action to combat climate change and its impacts[1]).

As a first requirement, governments have to take their inspiration from objective facts and figures derived from peer reviewed scientific papers. The UN sponsored IPCC has developed a reputation of seriousness and independence that is firmly established. Governments must develop their evidence based strategic plans drawing on the most recent IPCC reports.

In addition, governments have to develop the capacity to both listen to and anticipate demands from the street and social media. The 'noise' around climate issues is immense, confusing and often conflicting. It is important not only to feel the pulse of those concerned by the new threats created by climate change but also to be alerted by the claims and worries of the rest of the population. Change is the fundamental message but change is threatening, particularly to those least able to adapt; the less educated, those in later life and those with little resources. A proper understanding of the diversity of social demands is a must for properly designing political responses. The necessary leadership will require governments to assume the difficult political responsibility of drawing up a climate plan that identifies, filters and evaluates public concerns of all types. An obligation for all political leaders to undertake specific training on climate issues, mitigation and adaptation should be taken at all institutional levels in Europe, as recently proposed by the French association Comité 21 in a report devoted to adaptation.[2] Climate plans must integrate a regular process of evaluation by civil society to correct or re-orient action whenever necessary.

Many of the action points of the manifesto imply re-legitimising the role of the state, which has been disputed and downgraded since the turn of the 1980s to better reflect free-market economic principles. A renaissance of public intervention is required to get better prepared for tomorrow, This is amply justified by the complexity and far-reaching nature of today's challenges: climate change and the necessity to carry out the Transition in a fair way, but also the rise of inequalities and poverty, regional imbalances, the growing gap between urban and rural areas, new social demands and the spreading social and environmental damage caused by a public governance narrowly centred on economic returns.

[1] https://sustainabledevelopment.un.org/sdg13. Accessed 3 June 2020.

[2] Dayan and Simonet (2020).

The state's role cannot be limited to seeking to restrict or compensate for the damage caused by the excesses of the free-market economy. Re-legitimising the role of the state means concretely that it has to be entrusted with the responsibility to design and implement a new social contract built upon an environmental economic model and made up of ambitious, inclusive and voluntarist policies. These policies should be designed to optimise responses to the intertwined issues to which the whole of society is exposed, with climate change placed centrally.

Very often governments have a truncated view of their responsibility in managing the Transition. They focus their entire attention on creating the conditions in which low and zero carbon technologies, goods and services will prosper on the market. In doing so, they omit an essential aspect of the Transition, namely the setting of proper conditions for speeding up the phasing out of unsustainable practices. In restricting or abolishing unsustainable practices through regulation this must also be accompanied by support measures aimed at helping the economic actors to restructure or redirect their activities. The extensive set of interventions that were taken by governments to deal with the COVID-19 pandemic will undoubtedly broaden the scope and improve the perceived legitimacy of governments in taking such actions.

In many countries, re-building trust and confidence in the state will not be easy. However, there are areas where it is increasingly recognised that firmer government action is essential. Demands for combatting the excesses of the financial system are on a steep rise and are getting wider and wider support across society. The invasion of privacy and the use of personal data for commercial and political ends are also trends where state action is being increasingly called for. Each time, such calls are the expression of a society in need of political responses. According to the French philosopher Frédéric Worms,[3] the question of fixing limits is probably the most important question of today. The responsibility of the state is to redefine progress by fixing new limits: scientific limits, political limits, ethical limits, and even possibly limits to some unsustainable economic expectations driven by the excesses of capitalism. To have limits to carbon emissions, pollution, wastage or even to indifference vis-à-vis others must be conceptualised and communicated as progress to the whole of society. This will involve a process of regulation and entrusting existing or new institutions to ensure and enforce respect for the limits. For this to be received as progress by society, the control of limits must help to reconcile the double aspiration for a secure and inclusive society on the one hand and for a society that re-humanises and emancipates on the other hand. Scepticism will abound and there is no doubt that politicians will need to concentrate on cross-party approaches and take a new approach.

Mitigating climate change is the perfect example of an issue requiring the fixation of limits. Many such limits have already been partly set (e.g. to reduce the carbon emissions of the transport sector) but they will have to be augmented to sustain the more painful, future carbon reduction efforts. The manifold dimension of climate change will also require the imposition of new limits in domains that have been hardly regulated so far, such as the right to consume very polluting goods or services. The

[3] Worms (2019).

barriers to sustainable consumption are many and diverse, which calls for a certain form of planning or obligation to guide behaviour in a more predictable way. Limits on consumption and their control must be well understood and accepted to stand a chance of receiving social legitimisation. There is obviously no question of aiming at a green dictatorship (as proposed by Agnès Heller[4] and Hans Jonas[5]). However, a realistic, fearless and challenging reflection is needed to find ways to regulate those polluting activities that cannot be self-mitigated by the market or by spontaneously emerging green attitudes. The focal point of such regulation should be the reduction of the consumption of non-vital goods and services whose production is very carbon intensive. The goal must be, as phrased by Razmig Keuchevan, to imagine a form of governmental planning that is at the same time democratic and environmentally sustainable.

Governments have also to respond to another huge challenge: finding ways to have a large majority of the society acting in the same pro-climate direction. Companies, regional and local authorities, diverse organisations, financial investors and the general public must all be on the bridge, seeking to steer in the same direction. All these groupings, often referred to as "non-state and subnational actors" under the Paris Agreement,[6] can be considered as on-the-ground implementers of the accord. Governments acting as leaders of climate protection will have to create strong alliances with these non-state actors. At the local and regional levels, subnational authorities are often more active and engaged than the upper national governance level. For acting more decisively in coalitions with local communities, they will have to get more support from central government, notably to enable civil society to become the main driver of change.

As we move away from the narrow idea of 'greening' to the concept of 'transformation' and as the most innovative transformation projects often originate and flourish at a local level, there should be growing recognition of the need to actively support sub-national authorities and civil society. However, most national governments still seem to be stuck in the old 'greening' paradigm involving the belief that we can continue with the current economic model modifying it superficially but with little systemic change. Leadership by central governments does also mean that they massively encourage the local authorities, a(communities and neighbourhoods to implement the Transition everywhere, urban or rural. At stake is a shift away from the centralised, corporate model, towards a more localised model made up of a higher proportion of small local business and co-operatives. The governments' efforts to support a just Transition will also be best implemented at the local level where more inclusion and a better sharing of wealth to create opportunities are the main drivers of community projects in various domains (community energy, addressing energy poverty, community food production, etc.).

[4]Heller (1976).

[5]Jonas (1984).

[6]The reason being that they are not signatories to the Agreement.

In the EU, many proposals from the European Commission, strong in their original concept and evidence based, are made less effective by adjustment for national interest and by the pressure from lobbyists. Examples are numerous, e.g. energy efficient building standards.

Nevertheless, the reality is that European public opinion is a long way from being united on the climate issue. Wide variation in affluence between and within states, differences in the strength and the coherence of civil society are certainly factors in this divergence but the influence of national governments in their way of communicating the climate issue to the people is also important. Although there are many different reasons why people are reluctant to act, governments have an important responsibility in orienting public opinion, as suggested in the second action point of the manifesto. Strong governmental leadership must entail, where possible, capacity to mobilise those who are least convinced or if this cannot be achieved to offer reassurance that their interests will not be forgotten.

National governments have limited means to implement the required level of intervention, not least from a financial point of view. Central governments are struggling to combat the excessive power of large multinationals, especially in the technology sector, yet their co-operation and resources will be essential in resolving the climate crisis. National governments have therefore another kind of obligation to meet: developing concerted approaches with other countries, and not only the neighbouring ones.

Multilateralism is in retreat just when collective action is desperately needed to tackle complex and interconnected global challenges. Even in the EU with its established mechanisms of dialogue and cooperation the response to the covid crisis has been defined mostly from a purely national perspective with little synergy and joint preparedness, in part because health is not a common EU policy but principally because the national level is erroneously considered as the only level of action. Governments' capacity to lead has to be measured by their capacity to positively influence international climate governance at both the UN level (i.e. COP process) and at regional and bilateral levels. The EU's asset is its long history of inter-country cooperation in a wide range of policy domains. Even if the EU reaches actual zero carbon by 2050, it would be insignificant unless China, India, and the USA are on a similar trajectory. A major effort to reach net zero carbon by the EU, as targeted by the European Green Deal, will have to be accompanied by a similarly challenging effort to mobilise the entire international community and in particular the other major GHG producers. Meeting the Paris Agreement requires a joint global action. There is an abundance of research proving that by acting together, governments can save both money and time. There are many practical reasons why countries must cooperate between themselves: to share technological deployment and development, co-finance large projects, interconnect networks and facilities or run joint pro-climate campaigns.

Last but not least, strong climate governance must be accompanied by a clear framework with targets, indicators and a robust monitoring mechanism. Governments have to equip themselves with operational tools to keep a regular overview of the progress made at different sectoral levels. A climate policy scoreboard would enable them to address challenges on time and promote policy alignment at different levels of governance.

Turning Climate Protection into a Consensual Transformation of Society

In reflecting on the nature of responses to climate change, we are systematically brought back to the necessity of transforming today's small, minority movement into a majority one and then into embedded social and economic attitudes and actions. Achieving this kind of radical transformation demands the creation of a new, inspirational narrative. The grand political utopias of the 60s and 70s have lost their power of attraction; can something new be created to fill society with hope and a positive perspective? The idea of progress has to be collectively reinvented and there should be much more than mere technological innovation or endless consumption involved.

This new narrative is necessary to combat despair, the most powerful inhibitor to action. Despair is not rational and, in relation to climate change, is principally caused by an overly negative public perception of what is involved in acting for climate protection. In the minds of people, very little can be done to counter the dramatic climate trends that are seen as almost inevitable. Moreover, what could be done is often understood as leading to a reduction of essential liberties such as consuming or travelling. Such is the demotivating power of this perceived inability to make a difference many climate sceptic groups encourage public despair alongside their frequently used weapon of fake news.

But the situation is not hopeless. Acting does not mean regressing socially or economically. Rather, undertaking positive climate action is a factor of social and cultural emancipation that offers great potential for economic return in the medium term. If the whole of society could create a massive bottomup movement of positive climate action, we would, at a certain point, see actual improvements, notably in the form of the reduction of sectoral carbon emissions or even a slow-down in the temperature increase. This turning point, though still some years in the future, will provide huge motivation to continue on the pro-climate path.

Giving a positive, popular and contemporary image to climate action should become the prime responsibility of governments. To give substance to the image and to encourage action, a narrative made up of desirable and realistic objectives leading to a positive outcome needs to be framed. It has even to be considered as a first, essential step in boosting positive climate action across the whole population. But this first step is far from being straightforward. It is facing governments with the huge challenge of reversing the scepticism of the non-acting majority.

So far the twenty-first century has suffered from a shortage of mobilising narratives, especially in Western Europe where citizens feed their pessimism with a deficit of hope and fear of the future. A big communication campaign to motivate positive climate action has to be initiated without delay and must be developed on a medium-term perspective of a minimum of ten years. This communication campaign cannot be limited to a mere institutional, top-down approach. It has to rapidly become a multi-level, bottom-up endeavour where everyone shares their positive experiences through social media or within local communities. It must counteract the negative images that proliferate, prompted by those seeking to undermine positive action. Negative messages always benefit from larger audiences. The viewpoints of climate sceptics are featured in 49% more media articles than the ones of scientists recognising the anthropogenic cause of climate change,[7] in contrast with the growing percentage of the population acknowledging that climate change is happening (73% of US citizens in 2018 compared to 57% in 2010[8]). Climate change is a subject on which lobbying, irrational reactions and fake news have a very strong grip. The new environmental narrative must therefore attract the whole of society into a positive social and environmental transformation that gives back meaning to everyone's life.

Changing hearts and minds requires a profound change in communication with the general public. For different reasons, politicians have, in a majority of European countries, lost their capacity to inspire society with uplifting values. The most engaged politicians are at best able to influence a limited audience. This is particularly the case of Environment Ministers who are often occupying second rank responsibilities in their governments.

The message of asking people to do something for the climate is not easy to convey. People will only mobilise if they find an interest for themselves and if they develop the conviction that their action makes a difference. With around 60% of the EU GDP relating to domestic consumption, people have to be told that every purchase they make or refrain from can have an impact, either negative or positive. An important driver of people's action is its framing into a stylish, popular and positive collective endeavour. The communication campaign must be carefully sketched out to steer and entertain. Frightening or negative words and slogans must be used sparingly to maintain good will. Even though the use of the phrases 'climate emergency' or 'climate crisis' are valid in depicting recent trends, they can act as a deterrent to action.

The challenge of the communication campaign is to break barriers, overcome inertia and unite individual actions and good will to create a truly transformative social movement. The diversity of civil society calls for both different messengers and different forms of communication. If the younger generations are ready to act for the climate, the older ones are not always ready to question their lifestyle. The generational gap is even larger with the Baby Boomers who are reluctant to reform a system which gave them opportunity, placed them in important managerial positions in public administrations or at corporate level and rewarded them with pensions.

[7]Petersen et al. (2019).

[8]Greco C, What's Wrong With the Way We Communicate Climate Change?, Environmental and energy study institute (EESI), 24 May 2019.

Clearly, the communication has to be fine-tuned to reach out to the wide diversity of cultural and generational backgrounds.

The task of communicating should be shared between "climate action ambassadors", i.e. influencing personalities who are able to speak in clear terms of what it means to be doing something positive for the climate whilst having charisma and benefitting from a high public profile and audience. Typically, these ambassadors could be actors, singers, footballers, faith leaders, writers or philosophers. They should be able to influence segments of the population that cannot be reached by politicians. Governments should share the responsibility of mobilising the country's population through apolitical cohorts of well-known personalities. In the same way as web influencers, these personalities would address different groups of the population so that the whole of society would have at least one or two advocates addressing them in their own cultural language.

A grouping of French artists and personalities has recently created the collective The Freaks to train their fans and followers to adopt climate friendly gestures.[9] A list of realisable actions and their positive consequences has been prepared with experts, in particular from the Nicolas Hulot Foundation for Nature and Humans. The artists pledged themselves to adopt the proposed actions with the aim of better promoting them to their respective audiences. In all European countries, there exist such influencers who could convey strong messages as part of a large communication campaign. Climate action's ambassadors ought to stress that true modernity involves challenging one's own habits when it comes to protecting our common future. The decision taken by the pop group Coldplay in December 2019 to pause touring until they could offer environmentally neutral concerts is another example of the positive influence that artists could have on their fans and the whole entertainment sector.

In such a cross-cultural campaign it is very important that messages are built on a common set of principles and the many ways of acting positively for the climate are consistent and available to every sector in society. Providing a core set of coherent messages should be the responsibility of governments through a process of deliberative and consultative engagement with professionals, experts and the public.

As for the forms of communication, they must also tap into the great diversity of traditional and social media. The latter was initially shunned by politicians and elites but recently social media has been perceived as a way of reaching a target audience with a political message, unfiltered (and therefore unchallenged) by conventional media channels. Climate action ambassadors must responsibly use all kinds of social media, some of them having already demonstrated their capacity for raising awareness and funds to support pro-climate measures. Naturally, the process of using social media channels will also create negative and, almost certainly, aggressive reactions but this has to be seen as part of the necessary debate. All forms of communication are to be used and politicians themselves need to improve their communication skills through social media, in support of the role of climate actions' ambassadors. The internet is both a metaphor for and a facilitator of today's more horizontally structured society, made up of individuals with very different backgrounds. As a

[9] https://www.the-freaks.fr/collectif/. Accessed 3 June 2020.

result, communication must be both multi-faceted and very specific to address the full spectrum of profiles.

Last but not least, to prevent distorted and inconsistent messages to be passed on to civil society, it is essential that the communication package on the collective benefits of acting for the climate is largely supported by the main political parties, from left to right. A high-level political consensus is indeed essential to prevent the communication to be jeopardised by partisan interests. This kind of cross-party cooperation is rare in the history of democracies and has principally happened in very difficult times such as wars or periods of severe national crisis. Today's times cannot be compared to a war but they are sufficiently threatening to justify political unity. It is absolutely essential that all political parties can understand the vital importance of creating a seamless coalition of interests around the question of climate action. This might be wishful thinking but attempting it will, at the minimum, show to the population the supreme importance of acting together for the climate.

If this massive communication campaign involving European citizens is to deliver the expected outcome, it should receive the widest backing and support from public authorities and institutions. Similarly, diverse organisations from across civil society should also help to incentivise national awareness with consistent and positive messages. The recent decision of the British newspaper, The Guardian, to refuse advertising from fossil companies provides clarity and consistency with their editorial line which for years has been vigorously pro-climate. The newspaper recognised that it would be consistent, but not yet economically feasible, to extend a similar approach to car manufacturers and their transparency in this respect brings a breath of fresh air that is potentially encouraging for the uncertain and undecided. It is nevertheless illusory to believe that a perfect alignment of the message on climate action will happen overnight across all organisations. The role of the climate action's ambassadors is also to explore areas of tension, conflict and uncertainty and through dialogue shape a series of messages that value the role of everyone in climate protection.

At the EU level, the forthcoming Climate Pact is creating great expectations as the participatory arm of the EGD. Announced in autumn 2020, part of its challenge will be to turn the traditional approaches aimed at enhancing citizens' information and participation in EU institutional processes into more modern deliberative approaches that can place European citizens at the centre of the Transition.

Partnering with Civil Society and Citizens to Develop and Implement Policies

Acting for the climate must mobilise all the forces in society: people, public authorities, big and small companies, farms, local communities, non-governmental organisations, and religious organisation—all of need to share a common purpose. The role of the private sector is so critical that it is considered in the action point 6 on mobilising the business world.

There is a strong demand of the recently-created youth movements to directly involve young people in all discussions to fix the climate. Generation Climate Europe has in particular demanded to be part of the preparation of every aspect of the EGD. The 2030 SD Agenda acknowledges young people as "critical agents of change",[10] hence the necessity to give specific space for them to actively participate in the co-elaboration of climate policies. This particular strength and capacity of young people to get engaged is a key to implementing the necessary bottom-up transformations.

Individuals should act not only as responsible citizens but also as employees at work, as members of sport clubs or as consumers of goods and services. This idea is not new. Alexis de Tocqueville[11] noted that, "the collective strength of citizens will always be more powerful for producing social well-being than the authority of government". Now we have the means to reach every member of society directly—it has never been so easy to convey a message. The challenge is to convince everyone that the message is not only for them but that a sustainable future depends on their response.

Individuals can do a lot but if their initiatives remain isolated then reversing current emission trends and environmental degradation will be mission impossible. If, however, individual initiatives can be multiplied and brought together under a new, positive narrative, then the potential is massive, not only in terms of reducing impacts but also in terms of enhancing cohesion in society to deal with the risks that lie ahead.

Religious organisations should be involved because they have a large captive audience and ready-made forums for discussion and mechanisms for joint action. For example in February 2020 the governing body of the Church of England which has some 16,000 large buildings under its control, pledged to reach net zero carbon emissions by 2030—15 years earlier than initially planned.

New social innovations are proliferating in sectors as varied as food, energy, the circular economy or transport. Alongside the powerful driving forces of our market economy the creativity of projects originated by people and communities, without top-down constraints or profit requirement, is generally very high. Boosted by fierce enthusiasm and pride, these social innovation projects do, nevertheless, often require some financial support or greater visibility to deliver more meaningful outcomes. Local government should support these initiatives and not regard them as competing with their own. Removing the obstacles to their diffusion, accelerating the exchange of good practices or providing the necessary expertise for their further expansion, are the kinds of measures that could boost the confidence of people to do more and better for climate protection.

There is also another form of people's involvement in climate protection that is equally important, namely their participation in the making of public decisions. In today's society, flooded by ever-expanding fluxes of data, citizens show a growing

---

[10] https://www.undp.org/content/dam/undp/library/Democratic%20Governance/Youth/Critical%20Agents%20of%20Change%20-%20Youth-Inclusive%20Governance%20Indicators%20for%20National-Level%20SDG%20Monitoring.pdf. Accessed 3 June 2020.

[11] de Tocqueville (2002).

appetite for more meaningful information and participation in any kind of public decisions. This bottom-up demand results from the rapidly rising dissatisfaction of people with the functioning of democracies and their inability to meet the most pressing demands. The current European Commission has recognised the need to adjust our democratic model to this reality by having one of its six priorities dedicated to "A new push for European democracy".[12] In the course of 2020, it will adopt a European Democracy Action Plan which should help to counter disinformation and to adapt to evolving threats and manipulations, matters that are of great relevance for climate action.

The growing appetite for public participation in political decisions is confronting governments with what Michel Foucault called in the 1970s a "crisis of governmentality" (the word was coined to define the exercising of power and its questioning by the governed). Today, according to Dr. Jean-Claude Monod, Research Director at CNRS,[13] a new crisis of governmentality is caused by the supremacy of the economy in society over the environment. The challenging of political power by the people is now principally fed by social inequalities on the one hand and the inaction of governments vis-à-vis climate change on the other hand. In both cases, according to Jean-Claude Monod, some citizens dispute the adequacy of the politics of governance of the social and environmental issues. The voice of the people in this group expresses itself as: "we do not want to be governed in this way" and sometimes as "we do not want to be governed at all".[14]

The limitations of universal suffrage, often erected as the lighthouse of democracy, have never been more apparent than today when states and governments are mired in dealing with the first-of-a-kind challenge of climate change. By itself, universal suffrage is not able to guarantee a democratisation of the political decision-making process. The will of the majority is not enough to define democracy. The representativeness of those elected through the voting process is increasingly disputed. In Belgium, one fourth of the population is "very educated" whereas this percentage goes up to 93% in the federal Parliament.[15] Of course, being a representative of a group does not necessarily involve being demographically similar but there is a danger that those who govern may be seen as having joined an exclusive group.

Finding practical ways to mitigate climate change will require much more mutual understanding and consensus than for other routine kind of decisions. The competition between political parties which is inherent to the electoral process, prevents such a wide consensus from easily emerging. If a party wants to win extra seats, it has to capture these seats from other parties, hence fuelling antagonism. This competitive approach goes against consensus-building and the construction of a collective vision, a much needed goal when addressing climate change.

---

[12]Commission work programme 2020—A Union that strives for more, COM (2020) 37 final, 29/01/2020.

[13]Monod (2019).

[14]Interview in the French newspaper Libération, 12–13 October 2019, pp. 24–25.

[15]Interview with David Van Reybrouck, Le Vif, 28 November 2019, pp. 28–29 (in French).

The legitimacy of a policy does not measure itself only by the validity of the decisions taken but also by the way these decisions have been taken. For moving from a right to vote to a right to give opinions on policy projects, democracies have to be reformed to integrate new participatory mechanisms. A reform does not mean a full replacement. One should not make the mistake of throwing out current democratic practices with the bathwater. Instead, new, complementary approaches to the ones already in place should be developed, aiming at a representative democracy that would embed structural participatory and deliberative processes. Direct democracy requires a comprehensive supporting framework of education and deliberation which is probably why no national examples exist.

A majority of European countries have tried out different ways to supplement universal suffrage: referendums, popular consultation, right of petition, etc. The Swiss model, with its direct instruments such as popular initiative and optional referenda, shows that it is feasible for citizens to engage with complex questions.[16] The Irish government put in place a Constitutional Convention (2012–14) and the subsequent Citizens' Assembly (2016–18) tasked with providing good ideas on contentious issues such as same-sex marriage, abortion, the length of the presidential term of office, the voting age and climate change. This innovative participatory process helped to demine the question of abortion legalisation, with the ensuing approved referenda on same-sex marriage in 2015 and a constitutional ban on abortion was overturned in 2018. The Assembly also recommended strong climate action, including a high carbon tax.[17] All these practices aim at providing space for people to express their opinions and participate in decisions between elections. Experience so far has shown that citizens' assemblies, either self-created or created as part of a political process, are often very consensual, despite the great diversity of interests, class and status. Deliberation seems to generate greater awareness of long-term problems and makes participants more responsive to the interests of future generations.[18] The challenge, however, is to replicate that process of deliberation leading to consensus in the population as a whole.

The recommendation of David Van Reybrouck to proceed to the constitution of citizens' assemblies by drawing lots[19] in the Athenian way, appears as a very effective way to guarantee a representation of diverse lines of thought When a large number of citizens drawn by lot do respond positively, there is even the possibility of realising a second selection process to guarantee representativeness on the basis of criteria such as education or socio-economic status. These two criteria are key to properly embracing the variety of thoughts related to climate and social justice issues. Drawing

[16]The Swiss people massively support representative democracy (87%) and are less enthusiastic about direct democracy (67%), according to New Global Survey—Democracies under pressure—volumes 1, 2 & 3, FONDAPOL, May 2019; http://www.fondapol.org/en/etudes-en/new-global-survey-democracies-under-pressure-volume-i-the-issues/.

[17]New Global Survey—Democracies under pressure—volumes 1, 2 & 3, FONDAPOL, May 2019; http://www.fondapol.org/en/etudes-en/new-global-survey-democracies-under-pressure-volume-i-the-issues/.

[18]Mackenzie (2018).

[19]Van Reybrouck (2018).

by lot is also a way to prevent domination of the deliberative processes those willing to be most active.

Selection by lot is the way by which two very innovative participatory approaches have been recently set up: the creation of a permanent citizens' assembly representing all strata of society by the German speaking community of Belgium and the carrying out of a citizens' convention on climate by the French government.

The new experience of deliberative democracy set up by the Parliament of the German Community of Belgium in mid-2019 is illustrative of what can be done to create a larger space of co-decision between politicians and citizens. Constituted by 24 citizens drawn by lot, the newly established Citizens' Council is a structural body that has the right to decide topics of investigation on its own. Each topic will be debated with the aim of leading to concrete policy recommendations. This can be seen as a quasirevolution in the democratic processes of the small German Community of Belgium (only 75,000 inhabitants). Because of this Citizens' Council, politicians will lose the often-used capacity to sweep annoying dossiers under the carpet. Furthermore, they will be legally obliged to respond to the concrete proposals made by the Council. Rejecting the proposals will not go easily, as citizens will want good justification. This new Belgian model was developed in cooperation with G1000,[20] a platform for democratic innovation in Belgium. It goes much further than similar initiatives carried out in other countries or cities (e.g. Gdansk) in so far as it is a permanent process that is almost fully in the hands of the involved citizens. For the time being, the climate question has not been referred by the Citizens' Council but it should not be too long before it is put on the table.

Another sizeable experience of co-producing policy content has taken place in France in the second half of 2019 under the Citizens' Convention on Climate umbrella.[21] Presented by President Macron as one of the responses to the yellow vest movement, the convention brings together 150 French citizens who have been randomly and regionally selected to come up with proposals to feed the preparation of a major climate law by mid-2020. President Macron promised that the proposals issuing from the convention would be submitted, unfiltered, to either the French Assembly or to a referendum.

Similarly, in the UK, a climate assembly has been recently established by the government. 'Climate Assembly UK: the path to net zero' will consider how the UK can meet the Government's legally binding target to reduce greenhouse gas emissions to net zero by 2050. The first of several deliberative sessions involving 110 people took place in January 2020. An initial group of 30,000 households were chosen at random and responders were selected to be representative of society with a mix of ages, genders, ethnic backgrounds and education levels, but also a range of views about the climate crisis. Some of the questions debated included "Which is better for the environment—British beef or an avocado from Peru?", "What do you think should be the balance of business and government action on climate change?", "How committed are other countries to net zero?" and, "Is there an argument for letting climate change happen?".

---

[20] http://www.g1000.org/en/. Accessed 3 June 2020.

[21] https://www.conventioncitoyennepourleclimat.fr/en/. Accessed 3 June 2020.

For all these promising initiatives the final outcome will critically depend on the way the citizens' recommendations will be used by politicians. A partial or, even worse, full dismissal of the recommendations will probably signify the end of similar practices in these two countries, with also very negative consequences in the rest of Europe. This not only would be bad for democracy but also for the principle of combining forces to win big battles like the climate change one. It could also be seen as another boost to populism, a movement which eschews constructive, open debate in favour of preconceived positions.

Multiplying the routes through which the public can express its views is essential. There is no point in mobilising and engaging the whole of society with a positive agenda if people cannot feel part of the decision-making process. The fight against climate change must become a shared endeavour that is internalised and implemented by all actors and classes of society. It cannot be limited to only the ones who are engaged in the Youth for Climate, Zero Hour or other similar civic movements, as they only represent a minority of the population (though growing). All the deliberative processes carried out around the climate issue should be communicated and made visible to the entire population to enhance their social impact. For politicians, having citizens' assemblies proposing "difficult measures" could be a feather in their cap. This could help them to govern by taking fewer political risks and improving their chances to be re-elected while giving fresh fuel to a democracy running out of steam. Furthermore, it is also a means to prevent the holding of referendums which often have the negative consequence of polarising a country.

In the medium term, the objective must be to see the use of a wide range of diverse and locally relevant participatory means become the norm in a renovated institutional process. Deliberations taking place in citizens' assemblies, citizens' juries, deliberative polls, planning cells or consensus conferences must take place in "mini-public" sessions (with a mix of elected and governed people) with expert advice and moderation as a function of specific needs. An interesting example has been created in the last few COP meetings where several country delegations were made up of an enlarged group of representatives holding a very wide set of interests and opportunities. The more diverse the country representation, the more creative and acceptable its vision of the common interest. At a national level, the establishment of climate assemblies has been one of the key demands of some climate youth movements. Increasing the biodiversity of the stakeholders' base responsible for conceiving future decisions is an asset for the whole of society. Bringing together the representative and participatory democratic approaches, to complement each other, cannot but lead to more widely endorsed decisions. Among the different forms of citizens' involvement, the ones that make them work on concrete projects are to be favoured by governments as they always benefit from a higher level of motivation at the participants' level.

As reported by the EP study on the Global Trendometer 2019,[22] these participatory means could be backed by the use of digital technologies to facilitate exchanges and

[22]Global trendometer 2019—Essays on medium- and long-term global trends, European Parliamentary Research Service (EPRS) Study, ed. Eamonn Noonan, December 2019.

interactions between citizens and the decision-makers and to align bottom-up and top-down processes. The report cautiously points out that there should not be excessive expectations about "e-democracy". The e-participation platforms that have been put in place by some European countries have not always performed optimally because of issues of extreme polarisation, dissemination of misinformation and manipulation. The problems encountered, which are typical of online forums, can often be resolved by robust moderation. If and when resorting to e-platforms, politicians have to ensure their adequate moderation and put in place measures to correct their inherent flaws: problems of power and domination, risk of instrumentalisation by external actors, excessive time and resource demand etc. What is at stake is the advent of a form of European Democracy 2.0 for climate protection.

Innovations in democratic governance must concentrate on the quality of engagement, hence the need to combine e-participation with face-to-face deliberations. Recent experience shows that citizens have the capacity to engage in productive deliberation, especially when facilitated by expert support.[23]

The pressing task of governments should be to create incentives for multiplying effective deliberative processes to enhance societal convergence on climate solutions and sharing responsibilities. Deliberative processes must notably be designed to appeal to young people and prevent them from turning further away from democracy[24]; as emphasised by the Fondapol survey, there seems to be a decline in the importance attributed to voting by young people in general.[25] Climate protection requires an ongoing conversation with all parts of civil society and all citizens, young and older, to prevent accentuation of the current generational gap. This conversation must provide an element of continuity and stability in the way the climate issues are communicated to the public and addressed by policymakers between electoral cycles. This is a precondition for turning this global challenge into a positive vision that inspires and motivates.

Developing Capacity to Implement a Long-Term Vision

With the establishment of the UNFCCC, governments across the world set the long-term objective of stabilising "greenhouse gas concentrations in the atmosphere at a level that would prevent dangerous anthropogenic interference with the climate

[23]Democracy in the twenty-first century: towards new models of democratic governance, André Bächtiger and Claudia Landwehr, Items, 3 July 2018, https://items.ssrc.org/democracy-papers/democracy-in-the-twenty-first-century-toward-new-models-of-democratic-governance/.

[24]According to the Fundapol survey, 38% of young people (versus 24% for older cohorts) consider that "Other political systems might be just as good as the democratic system" and that "it is a good idea to have the government controlled by a strongman"; Fondation pour l'Innovation Politique, Democracies under pressure, a Global Survey, edited by D. Reynié.

[25]Fondation pour l'Innovation Politique, Democracies under pressure, a Global Survey, edited by D. Reynié.

system." The adoption of this objective has been a fundamental guide for the intergovernmental process and its Parties (i.e. the UN countries) in the fight against climate change. The Paris Agreement, reached at the 21st meeting of the Parties in 2015, is in itself a global strategy for the longer term that is principally defined by the temperature mitigation objective: i.e. to limit the average global temperature rise to well below 2 °C above pre-industrial levels and pursue efforts to limit this increase to 1.5 °C.

This goal provides a clear and strong direction of travel for the longer term. Achieving the long-term temperature goal would require global greenhouse gas emissions to peak by 2020 and subsequently be reduced to zero before the end of the century. To limit warming to 1.5 °C, this reduction to zero must take place around 2050. Achieving the objectives of the Paris Agreement therefore requires the establishment of a long-term strategy backed by a vision transcending political cycles and going beyond the specific interests of groups and individuals.

The problem is that we are living under the tyranny of the short term. Our liberal democracies are generally unable to give the future the attention it requires.[26] Al Gore said once: "The future whispers while the present shouts".[27] There are too many long-term problems that are not (well) addressed by democracies. Some decisions that are made or deferred today will significantly affect tomorrow, and not only in relation to climate change. Dealing with the retirement age and the related implications of an ageing society has been a typical example of a permanently pushed back decision.

Climate change obviously requires decisions to be taken now to reduce its long-term impact but one has to remain prudent in not taking hasty decisions that could be regretted later. In line with action point 3, long-term and complex issues require a highly consensual and deliberative approach to formulate decisions that could be largely endorsed by the general public. Enough time is also necessary to objectively assess the sustainability impact of different policy options. The use of biofuels from the first generation (made of crops) is a typical example of a decision which has not delivered the intended sustainability gain. Furthermore, an abrupt Transition could have destabilising effects on employment and on the financial system. Time has to be given to firms to put into place alternative strategies for a smooth adjustment that does not lead to a massive increase of stranded assets, workers or regions.

Expanding the time horizon of the decision-making process is hard to conceive for democratic governments that are geared towards the narrow horizon of the next elections. In the head of politicians, the long term often plays the role of a repellent as it is, by its very nature, unpredictable and cannot offer quick political gains to the ones defending its cause. Furthermore, long-term plans and programmes are not a tradition of democratic regimes where they suffer from the negative reputation of the central planning approach favoured by many former communist regimes.

[26] Thompson (2010).

[27] Gore (1992).

Developing a capacity to think and decide for the long term requires a profound re-setting of democratic processes and a restructuring of internal institutional agendas. The new long-term culture must be developed while keeping untouched the basic pillars of democratic systems such as regular elections and fixed-term government programmes.

Several innovative institutional mechanisms have been proposed to address the key shortcoming that future generations do not have a voice in the democratic process. To give representation to future generations is very hard to achieve. The mechanisms proposed by political scientists[28] are nevertheless far from being satisfactory because of the risk that the rights of future citizens are misused to trump the rights of current citizens.[29] There is no practical way to objectively substantiate the wishes of future generations. Nominating representatives for the future will always present the risk of offering biased perspectives based on today's ideas.

Yet with regard to climate change in particular, planning for the future can in many ways be modelled by taking the point of view of current generations' into account.[30] As demonstrated by the recent climate marches, current generations, or at least part of them, actually care about the future. By contrast, older generations are relatively less preoccupied by what is coming next, an attitude which lies at the heart of the considerable generational gap regarding treatment of the climate issue.

Discounting the long term is therefore more of an issue for elected officials than for their electorate. In other words, politicians wanting to prioritise policies which care for the future have to take inspiration from the views of current voters, and especially the ones from younger generations. Rama Iselin has concluded that "in order to protect future generations, we need to find better ways of representing the current generation".[31] She proposed focusing the current generation's attention on issues that matter for future generations instead of attempting to represent future generations. A trustee of the future and even a Superagency that watches out for the long-term could help to counter the biases that award a premium to present preoccupations.

Governments wanting to stimulate the take up of more long-term perspectives by current voters should limit the influence of money in politics as moneyed interests almost systematically orient politics towards present stakes. In the same way, deliberative democracy, as proposed in action point 3, could significantly help to express and explore long-term wishes. And in the same vein, enhancing governments' accountability could also help to create more public confidence for long term planning. Another advantage of operationalising long term strategies is that it enables a more progressive and balanced implementation of climate-related policies, creating less tensions between the young and old generations.

In the same way as the communication campaign described in action point 2, long-term plans have to be supported by a wide coalition of political interests in order to

---

[28]Thompson (2005).

[29]Iselin (2012).

[30]Iselin (2012).

[31]Iselin (2012).

ensure their regular implementation beyond the electoral terms. They also have to get precedence over the sectoral policies that are often decided outside the framework of climate planning and yet have a requirement to be consistent with overall climate policy goals. In order to guarantee that long-term plans are not scrapped by following governments, the latter should aim at contractualising the climate goals with public and private organisations. By giving a legal, contractual existence to climate mitigation measures, the risk to have measures taken by previous governments scrapped by the next will be significantly minimised.

Developing Integrated and Holistic Approaches to Underpinning the Longterm Vision

The challenges faced by liberal democracies are numerous and intertwined: risks of new financial or public health crises, rising inequalities, stagnating economic growth, loss of biodiversity, scarcity of natural resources, cyber criminality and of course climate change. A systemic and holistic approach is needed to identify and prioritise synergetic measures which can resolve various environmental, social and economic issues as part of an overall strategic package.

Fighting climate change can go hand in hand with rebalancing social and economic opportunities and reducing inequality. The increase in fuel taxation implemented by the French government in mid-2018 was the best counter-example of what governments must do in terms of policy coherence. Applied as a single measure, in an uncoordinated manner with other sectoral policies, it led to the yellow vest movement, a surprisingly long lasting and determined civic response. The movement catalysed action about a wider range of perceived problems but with climate change quickly fading from the overall picture.

The principal evidence of the inconsistency of governments' policies is that even the most climate ambitious countries continue to foster unbridled consumerism, their national industrial polluters or unequal economic liberalism besides their pro-climate policies. A truly holistic approach calls for a 360° renovation of the key policy foundations, and principally of the economic ones (as recommended in action point 9). A truly holistic approach necessitates a true level playing field between economic, social and environmental policies. None of this has even been envisaged by any country worldwide.

The policy framework which demands the greatest application of an integrated approach is the implementation of the 17 Sustainable Development Goals (SDGs) that compose the 2030 Sustainable Development Agenda which was adopted in September 2015.[32] The progress so far has been so meagre and uneven, both across and within countries, that world leaders called for a 'Decade of Action' when they

[32] Transforming our world: the 2030 Agenda for Sustainable Development, Resolution adopted by the UN General Assembly on 25 September 2015, https://www.un.org/ga/search/view_doc.asp?symbol=A/RES/70/1&Lang=E. Accessed 3 June 2020.

met at the 2019 SDG Summit. A 2019 OECD report[33] shows that, while there is progress in some SDGs, countries need to ramp up their efforts, especially to tackle inequalities and improve environmental sustainability, educational and employment outcomes, healthy behaviours and safety.

To accelerate action requires new ways of shaping policy proposals by looking at all sectoral impacts and optimising policy combinations. Governments must identify and reform the barriers to a climateneutral transition embedded within existing industrial strategy, trade policy, competition policy, innovation policy, labour policy and financial regulation. All policy domains must be aligned with the goal of achieving a climate neutral economy.

The SDG Summit Political Declaration,[34] adopted in September 2019, committed UN Member States to "proactively develop effective, accountable and transparent institutions at all levels and ensure more responsive, inclusive, participatory and representative decision-making processes….[to] strive to equip domestic institutions to better address interlinkages, synergies and trade-offs between the goals and targets through a whole-of-government approach that can bring about transformative change in governance and public policy and ensure policy coherence for sustainable development."

Addressing interactions between economic, social and environmental goals in a balanced manner, while avoiding negative effects on the wellbeing of people, has been recognised by many countries as one of the most difficult challenges to implementing the SDGs. The 2019 OECD Recommendation on Policy Coherence for Sustainable Development (PCSD)[35] calls for "fostering synergies, reducing trade-offs, addressing transboundary and inter-generational impacts, whilst promoting partnerships and collaboration."

Given that the Transition will affect many interests, it is essential to offer development opportunities to the poorest and the most exposed or, at the minimum, financial compensation. Policies and measures which ignore fairness and decency will be condemned to endlessly face the wrath of the population, perpetuating the yellow vest movement in different forms. The goal pursued by decision makers should be that all groups of society become winners in the Transition. For this to happen, governments have to be equipped with the capacity to steer more integrated and inclusive policy-making approaches that optimise collateral impacts of a given policy goal.

Maximising the returns of policy decisions requires systems thinking and holistic policy-making but until recently, systems thinking seems beyond the capacity of policy makers. Now the global impact of COVID-19 and the dramatic reduction in world trade and disruption of supply chains is likely to stimulate the Copernican

[33]OECD (2019).

[34]Political declaration of the high-level political forum on sustainable development convened under the auspices of the General Assembly, Resolution adopted by the General Assembly on 15 October 2019, https://undocs.org/en/A/RES/74/4. Accessed 3 June 2020.

[35]OECD Recommendation on Policy Coherence for Sustainable Development, Adopted on: 29/04/2010, Amended on: 11/12/2019, https://legalinstruments.oecd.org/en/instruments/OECD-LEGAL-0381. Accessed 3 June 2020.

revolution that is needed to change mind-sets and develop new collaborative, cross-department and cross-stakeholder, approaches. As stressed by Richard Heinberg,[36] governments are naturally inclined to approve those experts saying that "climate change can be framed as an isolated problem for which there is a technological solution". It is reassuring and allows "the minds of economists and policy makers to continue grazing in familiar pastures". Technology solutions, with their well-known development cycle (mobilising investments, setting tasks for engineers, managing the industrial transformation), have the advantage of not having to deal with the complex interactions between the earth and human systems. A mere technological approach is however largely insufficient to respond to the manifold dimensions of climate challenge.

To develop a capacity to counter the systemic and existential threat of climate change, governments have to enlarge the scope of policy proposals and combinations under consideration. Some authors use the neologism of "climatisation" of the decision-making,[37] to refer to the growing interconnection of the climate issue with the wide array of societal issues such as finance, land use or diet. Paradoxically, this necessary aggregation around the central climate issue makes it even more complex to design integrated solution-oriented policy packages given the large number of policies to agglomerate.

At the start of the policy integration process is the collection of viewpoints and claims. Governments should utilise scientific and experts' views in a more effective manner. They generally face more and more complex problems to handle, as was demonstrated by the COVID-19 pandemic, and the recourse to scientific expertise is often essential to efficiently address these problems and develop possible solutions.

The reality is that climate science is not optimally used in governmental policy-making, despite the model reputation acquired by the IPCC in the conduct of the work of the UNFCCC. It is well known that national policymakers do not always inform scientists about their needs for scientific expertise for elaborating new climate policy proposals, but there is another, more serious, reason for this shortcoming: the politicisation of climate science or, in other words, the manipulation of climate science for political gains. Quite manifest in the US from the 2000s, it has become more and more difficult for politicians to value the role of science in public debate and policy. Many corporate-funded campaigns to play up uncertainties in climate science, mostly carried out through industry associations, play their part in sowing public confusion about the level of consensus in climate science. And in recent years, it has even proved to be harder to connect climate science with politics because of the growth of populism and the change of the political environment caused by social media. Two out of three right-wing populist Members of the European Parliament (MEPs) regularly vote against climate and energy policy measures, according to a

[36]Heinberg (2017).

[37]Conference at Paris-Sorbonne University on the "climatisation" of the world, 10 June 2016, Paris, France, https://www.urd.org/en/event/conference-at-paris-sorbonne-university-on-the-climatisation-of-the-world/. Accessed 3 June 2020.

recent Adelphi study.[38] Populist party platforms are often hostile to policy designed to address climate change, placing their leaders and supporters outside the political and scientific mainstream. In Finland, the populist Finns Party accused mainstream politicians of "climate hysteria" and argued that environmental measures would "take the sausage from the mouths of labourers".[39]

Apart from science, governments have to develop streamlined approaches to gather the views of the different economic and social sectors in order to facilitate the emergence of solutions that could benefit the entirety of civil society. In doing so, they have to resist the heavy influence of lobbies and economic elites by setting new, stricter rules guaranteeing transparency and restraining advocacy within certain limits.

Last but not least, it is also essential to take on board the views of non-experts and of non-stakeholders (the ones with vested interest), in other words of laypersons. Fresh, non-expert eyes can bring much added value to explore new policy horizons in a taboo-free manner. The participatory and deliberative democracy methods highlighted in action point 3 would help to tap into the creative ferment of citizens' ideas.

There is a fair range of successful experiences and methodologies for applying integrated approaches in various domains. For instance, the Managing for Sustainable Development Impact (M4SDI) methodology developed by Cecile Kusters and Karen Batjes[40] is an integrated, results-oriented management approach, which can be used across a range of sectors and domains in a variety of contexts. With its people-centred approach, it can support governments in contributing towards the SDGs of the UN SD Agenda.

Reaching out effectively to various actors and good methodologies are not enough to successfully bring together the various policy issues influencing climate protection. A coordination at the highest governmental level is required to ensure that the different Ministries are acting cohesively. Many European Prime Ministers have created specific political and administrative structures to assume a cohesive approach towards the implementation of the 17 SDGs of the 2030 Sustainable Development Agenda. Existing structures could be used to apply the same type of consistent approaches in addressing the climate changes causes and consequences throughout the entire spectrum of ministerial responsibilities.

The new governmental commitment to think and act holistically has to be adequately communicated to civil society. The rhetoric of governments must refrain from saying that "we have to change everything, including our entire economic and social system". This kind of message, because of the many open questions that it raises, can be a killer for people's motivation. Instead, the emphasis should be placed on the many advantages that could be created by an ambitious and balanced climate

---

[38] Schaller S, Convenient truths, Mapping climate agendas of right-wing populist parties in Europe https://www.adelphi.de/en/publication/convenient-truths.

[39] Lockwood M, Right-Wing Populism and Climate Change Policy, https://www.oxfordresearchgroup.org.uk/blog/right-wing-populism-andclimate-change-policy. Accessed 3 June 2020.

[40] Kusters and Batjes (2017).

plan, stressing that the climate solutions could be profitable to all persons and organisations in society if all of us are united by a strong resolve. The Transition can be a lever to uplift social justice and reduce inequalities, this is the kind of message society needs to act cohesively.

Mobilising the Business World

Reforming the way the different levels of government take decisions and incorporate interrelated issues will not be enough if it is not accompanied by similar changes within public and private organisations. Governments and their administrations cannot be the only promoters of long-term, integrated and holistic approaches. In an ideal, cohesive society, all organisations, public and private, would have to align their plans and strategic approaches with the governmental vision of a decarbonised society. Shaping a desirable future, with and by strong public engagement, implies a role for all organisations, whatever their size and *raison d'être*. Organisations have to consider themselves as advocates of societal responsibility, namely, to care for the common good.

The case of the private sector's role is symptomatic of the kind of cultural blockage that too often constricts the extent of climate action. Public opinion tends to see the private sector as one of the main, if not the main, culprits of climate change. Though the historical responsibility of the manufacturing and energy production sectors is undisputed, there must be a more balanced vision of both the ongoing efforts of the private sector to reduce its climate footprint and, more importantly, the substantial role it could play in the overall years-long decarbonisation process. Taking the most optimistic scenarios regarding the engagement of individuals and communities in climate protection, the role of the private sector in meeting the Paris Agreement should be a minimum reduction in GHG emissions by 35%. Should individuals and communities be less motivated to act then the reduction by the private sector could be as high as −60%. This substantial reduction should principally be achieved through the decarbonisation of industry, energy, freight transport, agriculture, and public services.[41]

For over 200 years, governments have been shaping legislation to address health, social and environmental issues in the corporate sector.[42] Early legislation was often influenced by progressive and exemplary industrial leaders[43] and this dialogue has been essential for effective law-making and the regulation of the corporate drive for efficiency and profit maximisation, often while trying to externalise inconvenient costs. This continuing dialogue is essential in the effort to mitigate climate change.

---

[41]Dugast (2019).

[42]The 1801 Factory Act in the United Kingdom is one of the earliest examples.

[43]E.g. Robert Owen, a Welsh textile manufacturer and social reformer greatly influenced early British legislation on factory working conditions.

Meeting the Paris Agreement will require additional regulated and voluntary emission reductions by companies. Hard and soft legislation must be tailored made to stimulate more and better contribution by the private sector, in particular to incentivise those companies that are willing to get engaged in new sustainable economic models such as circular economy and sharing economy or that are at the forefront of climate friendly technology development. Legislation must also help to intensify companies' structural efforts to align their internal strategy with the Corporate Social Responsibility (CSR), Responsible Business Conduct (RBC) and Environmental, Social and Governance (ESG) principles. A further strand in this process is the evolution of business and human rights (BHR) which focuses on a more delineated commitment in the area of human rights.

The UN Environment Programme (UNEP), which has long had the private sector in its line of fire, is engaging it as never before to create a truly cohesive approach in bringing needed solutions, "recognising that the actions of governments and civil society alone are not enough for tackling our most pressing environmental challenges".[44] Following the withdrawal of the US government from the Paris Agreement, a series of initiatives involving the private sector with American states and the cities, such as the We are Still In[45] declaration, have quickly picked up the baton of climate action. Existing climate commitments and policies by cities, states and businesses alone would fulfil two-thirds of the United States' national emission-reduction pledge for 2025 which was made in 2015. More than half of the NDCs submitted under the Paris Agreement involved consultations with the private sector, but details on private sector contributions to the targets were not mentioned in submissions.

Except for companies engaged in climate technologies and services, the involvement of the private sector in the protection of the climate has been rather low, irregular and patchy so far. There are even companies, such as coal, steel, aluminium and cement, which have been stuck in high carbon-emitting processes for years without really attempting to reduce their footprint. Moreover, many of the large companies who have been using the climate as a marketing argument often failed to demonstrate a real contribution to its protection. Multinational corporations, in many instances, promote headline-grabbing actions while largely continuing business as usual. Thus, for example, Shell trumpeted plans to spend $1–2 billion per year on clean energy technologies out of a total budget of $25–$30 billion.[46]

For many exposed business activities, the big challenge of the years to come will be to enable fast transitioning to carbon neutrality without losing in the international competition battle. Positive indicators recently came from one of the most criticised business, namely the European aviation sector. As from November 2019, easyJet decided to offset the carbon emissions of the fuels used in their flights—the first major airline to do so. It will cost the company an annual 30 million euro. British Airways and Air France announced that they would carbon offset all of their domestic

[44]https://www.unenvironment.org/about-un-environment/private-sector-engagement. Accessed 3 June 2020.

[45]https://www.wearestillin.com/. Accessed 3 June 2020.

[46]Reuters 12.11.2018.

flights. In 2020 the German airline group Lufthansa has also launched a business fare where European flights are automatically offset for corporate customers. Even more reflective of the cultural change, KLM advises their customers in their *Fly responsibly* flyer to reflect before taking the plane, arguing that "railway or other modes of transportation can be more sustainable than flying, especially for short distances such as within Europe".

There are industrial groups that are in the vanguard of climate mainstreaming but these still constitute a minority. The Global Risks Reports of the World Economic Forum are an invitation to the private sector to look ahead across a broad range of issues increasingly dominated by climate and environmental risks.[47] The role of the World Business Council for Sustainable Development,[48] which is a global, CEOled organisation of over 200 leading businesses working together to accelerate the transition to a sustainable world, is a step in the right direction. Developing other similar networks at the international and national levels is an absolute necessity. It is also important that small and medium sized enterprises (SMEs) are using climate protection as a market opportunity while reducing the carbon footprint of their overall business.

There are very good reasons to argue that the business sector would be one of the largest beneficiaries of affirmative climate actions. The longevity of private enterprises depends on their capacity to reduce their reliance on expensive and rare primary resources and, at the same time, respond to the expectations of their customers. This requires innovative new production and consumption approaches that are better geared towards the long term and the preservation of finite material resources.

New business models are already beginning to transform linear processes of consumption and production on which enterprise was traditionally founded. One of the strongest drivers of the greening of business models in the EU has been the 2015 Action Plan for the Circular Economy,[49] which promoted the eco-design of production processes, the re-use, repairability and re-manufacturing of products, the market of secondary raw materials and, as a last resort, the recycling of raw materials. Many countries worldwide have similar plans in support of a circular economy. The ideas of a circular economy were officially adopted by China as early as 2002. In the US, nearly four in five companies have a strategic intent to implement a circular economy framework (62%) or have already put one in place (16%).[50] However the world economy remains only marginally circular. The First Circularity Gap Report by Circle Economy estimates that the world is circular by a meagre 9.1%,[51] with the EU having only 12% of its materials recycled and brought back into

[47]https://www.weforum.org/global-risks/reports. Accessed 3 June 2020.

[48]https://www.wbcsd.org/. Accessed 3 June 2020.

[49]Closing the loop—An EU action plan for the Circular Economy COM/2015/0614 final.

[50]Opportunity and disruption: How circular thinking could change US business models A circular economy survey, ING, 05/02/2019.

[51]First Circularity Gap Report, Circle Economy, https://www.legacy.circularity-gap.world/2018. Accessed 3 June 2020.

the economy.[52] Every effort to enhance circularity is a positive step forward in the fight against climate change. Circle Economy considers that "the circular transition provides actionable ways forward to contribute to reaching the SDGs and the Paris Agreement". It is therefore no surprise that the new EU Circular Economy Action Plan,[53] adopted in early 2020 by the European Commission, has become a key pillar of the European Green Deal aiming at climate neutrality. Half of total greenhouse gas emissions come from resource extraction and processing, meaning that achieving Europe's climate-neutrality target by 2050 will not be possible without transitioning to a fully circular economy.

The efforts of industry to become more circular will depend to a significant extent on the way consumers will themselves adapt or change their habits in the coming years. Many surveys and studies[54] show that people are willing to consume differently, stressing how important it is for corporates to be the drivers of a whole-of-society ripple effect towards the production and consumption of circular products and services.

As part of the developing new economic models, the collaborative or sharing economy also shows potential to increase the shared use of goods, improving their utilisation and reducing the need for new production. Usually defined as a peer-to-peer based sharing of access to goods and services, facilitated by a community-based online platform, the sharing economy provides a potential pathway to sustainable societies.[55] With some studies casting doubts on the sharing economy's environmental effectiveness, better governance models are urgently required to maximise the positive impacts on sustainability. In the transportation sector, vehicle sharing behaviour can have a positive environmental impact by decreasing the number of kilometers travelled and vehicles owned. Governments have a key role to play in identifying the sharing models that are the most pro-sustainability from a life cycle analysis (LCA) perspective and supporting service providers through both economic (e.g. lower taxes and subsidies) and non-economic incentives (e.g. communication campaigns and labelling).

A further transformation in corporate thinking must accompany these actions. Business needs to consider how it delivers its essential 'service' to customers in a climate-friendly way. In the energy sector the concept of energy services has already begun to supplant the concept of energy supply. This facilitates a move away from carbon-intensive practices. The business model of Energy Service Companies (ESCOs) is driven by energy efficiency projects that are financed based on energy

---

[52] https://www.euractiv.com/section/circular-economy/news/eu-unveils-circular-economy-plan-2-0-drawing-mixed-reactions/. Accessed 3 June 2020.

[53] A new Circular Economy Action Plan For a cleaner and more competitive Europe COM/2020/98 final.

[54] Behavioural Study on Consumers' Engagement in the Circular Economy, study commissioned by the European Commission, carried out by LE Europe, VVA Europe, Ipsos, ConPolicy, Trinomics, October 2018.

[55] Mi (2019).

savings.[56] Similar approaches are being applied in sectors like transport, housing, and communication. The vision of enterprises which render essential services to society must be expanded to all those contributing to a shared prosperity in sectors such as health, social care, education, culture or craft. This vision has been clearly articulated by Professor Tim Jackson: "An economy that works must have something to do with investing in work itself. Care, craft, culture, creativity: these sectors offer a new vision of enterprise: not as a speculative, profit-maximising, resource-intensive division of labour, but as a form of social organisation embedded in the community, working in harmony with nature to deliver the capabilities that allow us to prosper."[57] In addition to being leaner in material and energy requirements, enterprises of that kind contribute positively to the social pillar of sustainable development and the wellbeing economy.

The private sector is also best placed to bring new clean technologies to the market, a process where the EU is structurally less efficient than its main competitors like the USA and Japan. Though the overreliance on tech development has to be counterbalanced with a more diverse set of emission reduction measures, it remains essential to guarantee access to vital technologies to drive the decarbonisation of the transport, energy, building, and other sectors. New communication and information technologies have the potential to facilitate modern governance approaches resting on a closer cooperation with civil society. For local communities to become true actors in the transition, it is also necessary to develop new sustainable technologies that are adapted to their specific needs in terms of remoteness and size, an investment that has the potential to rejuvenate communities often left behind by years of underinvestment.

Historically, corporate social responsibility (CSR) has been the entry door through which some businesses have started to become active in strengthening the fabric of the common good. The three main pillars of CSR[58] are the three Ps—people, planet and profit—and are commonly translated as social, environmental and economic. The term CSR was coined in 1953 with the publication of Howard R. Bowen on Social Responsibilities of the Businessman.[59] It was increasingly implemented from 1990 by a growing number of large businesses. For some companies, it has evolved beyond mere codes of conduct and reporting and has started to become strategic, impacting all aspects of business' activity towards financial ends. In 2011, Michael Porter, Professor at Harvard University, went a bit further with his idea of 'creating shared value'.[60] The cause of the diminished trust in business has been their outdated approach to value creation narrowly focusing on short-term financial performance and ignoring the broader influence of evolving customer needs and negative societal

---

[56] Energy Service Companies (ESCOs)—At the hearth of innovative financing models for efficiency, IEA report, December 2018, https://www.iea.org/reports/energy-service-companies-escos-2. Accessed 3 June 2020.

[57] https://www.theguardian.com/commentisfree/2017/may/31/economy-neoliberalism-free-market-economics. Accessed 3 June 2020.

[58] Identified by John Elkington in 1994 as the triple bottom line.

[59] Bowen (1953).

[60] Porter M & Kramer M, How to reinvent capitalism and unleash a wave of innovation and growth, Harvard Business Review, January-February 2011.

impact. He called for the right kind of regulation to stimulate innovation and create a level playing field to invest in creating shared value instead of maximising short-term profits. Even if the term of creating shared value has not quite replaced CSR, the concept of "strategic CSR" has become largely embedded. Some will look at it as an example of neo-liberal capitalism seeking to defend its territory by adopting a more socially orientated approach whilst continuing with business as usual. Others will consider that climate urgency requires that all progressive actors work more closely together under the auspices of an ambitious and inclusive climate governance.

What companies do and how they do it have significant implications for the lives of citizens. These impacts are not just in terms of the products and services they offer or the jobs and opportunities they create, but also in terms of working conditions, human rights, health, the environment, innovation, education and training. Enterprise must be able to deliver the basic means for all citizens to meet their underlying needs for subsistence, nutrition, clean water, affordable energy, health and education. Also companies cannot continue to cast citizens almost exclusively as individual consumers. The injunction to consume has to be turned into a shared endeavour of deciding on sustainable, accessible and affordable alternatives, a process that implies a greater responsibility at both citizens' and companies' levels.

CSR or 'responsible business conduct' (RBC), an alternative term introduced by the OECD, imply that companies prevent, manage and mitigate any negative impact that they may cause within their global supply chain. The EU consumption has spill-over effects on the GHG emissions of countries which are EU trade partners. The Global Carbon Project,[61] a major effort to measure CO_2 emissions worldwide, attempts to measure this outsourcing effect by estimating the difference between the amount of CO_2 that each country produces within its borders (the official country's amount according to the UNFCCC) and the amount of CO_2 that is consumed by each country (taking account of the CO_2 emitted around the world for manufacturing the goods consumed by that country). Without surprise, wealthy OECD nations were "consuming" around 15% more than they actually emitted within their own borders in 2014.

The EU actively encourages companies to conduct their business responsibly, sometimes via mandatory actions. Recently concluded EU trade and investment agreements contain specific provisions committing the parties to promote CSR/RBC which refer to internationally agreed instruments in this area. For the European Commission, CSR is also regarded as a way to implement the UN guiding principles on business and human rights (UNGPs) and the UN 2030 sustainable development agenda.[62]

In recent years, many companies actively engaged in CSR have considered that the fabric of the common good has become a synonym for the protection of the

---

[61] https://www.globalcarbonproject.org/carbonbudget/. Accessed 3 June 2020.

[62] https://ec.europa.eu/growth/industry/sustainability/corporate-social-responsibility_en. Accessed 3 June 2020.

climate. Some have begun to redefine their CSR policies to prioritise the environmental component. This trend has been reflected in the growth of a new term redefining CSR priorities, namely ESG for Environmental, Social and Governance. ESG criteria are a set of standards for a company's operations that are increasingly used to qualify and quantify companies' efforts to reduce their contribution to climate change. They are increasingly used by conscious investors to screen potential investments. There is growing evidence that suggests that ESG factors, when integrated into investment analysis and portfolio construction, may also offer investors potential long-term performance advantages.[63] If the number of corporate announcements about large sums of money invested to steeply reduce carbon emissions or even bring them to zero, has been rising, it is in part due to the benefits derived from "positive" ESG factors.

Independent and objective control of companies' practices is indispensable to hold them to account to transform their good intentions into measurable improvement of their carbon footprint. CDP[64] is a notfor-profit charity that runs the global disclosure system for investors, companies, cities, states and regions to manage their environmental impacts. It provides the public monitoring, reporting and evaluation system for climate-related information of companies, cities, regions and financial institutions. This includes driving transparency, holding stakeholders accountable and bringing them together for knowledge and best-practice exchange. CDP's carbon disclosure rating does not necessarily reflect the actions taken by an organisation to reduce its impact on climate change and concentrates on the level of disclosure of information. Nevertheless such evaluation schemes are an essential first step in building robust accountability systems that can form the basis for statutory measures.

Another reason to have business in the partnership with governments and citizens is to prevent them from lobbying against change for their own short-term benefit, and to encourage them to promote the reality of climate change to their employees and consumers. Governments have to take measures to incentivise companies to assume more responsibility in climate protection, within and beyond their national operating base. Regulation and norms are obviously needed, but also the creation of a virtuous enabling framework where companies would, by themselves, anticipate further legislation. Many business networks, such as The Climate Group,[65] act as a catalyst to build a shared sense of ambition and pace through using the power of cooperation and communication. Governments could foster the creation of green business coalitions which would enable a larger number of companies to share knowledge on industry- or sector-wide challenges and opportunities relating to climate change. These coalitions could in return help government entities to seek input to develop sectoral climate policies and to identify existing market barriers. When the sectors are not responsive, the governments should create their own platforms to stimulate the private sector's engagement in climate action. Another route is to involve multinational companies that are demonstrating climate leadership internationally to share and encourage the

[63] https://paxworld.com/sustainable-investing/what-is-esg/. Accessed 3 June 2020.

[64] https://www.cdp.net/en. Accessed 3 June 2020.

[65] https://www.theclimategroup.org/project/business-actions. Accessed 3 June 2020.

take up of aspirational targets by other businesses. A private sector climate action data bank could also help governments to improve the design of their climate policies by getting access to a wide range of climate-friendly practices by different industry sectors. The reform of economic policies proposed in action point 9 would provide a strong stimulus for companies and their networks to get more engaged in developing products and services responding to well-being aspirations.

Workers and employees are key actors to trigger deeper consideration of accelerating carbon neutrality within companies. To ensure that businesses do really mainstream climate protection within their entire activity, their employees should be actively involved with the internal design of corporate approaches. Though there does not seem to be a correlation between the green nature of companies and the involvement of employees in their governance, it would seem natural that climate-responsible companies are leading by example regarding their employees' role.

The preservation and the upgrading of jobs along the transition must be backed by an enabling policy framework that gives recognition and stimulus to labour-intensive jobs in service-based activities. Combined with a lower environmental impact, the labour-intensive economy is the care and craft economy that is vitally needed to sustain the social, cultural and wellbeing dimensions of the transition. The inevitable downward trend of labour productivity and its accompanying downward pressure on wages will have to be compensated by appropriate policy measures.[66] Fostering the wellbeing economy and grasping the job opportunities of the future will require substantial investment in quality basic education, lifelong learning, up- and re-skilling and digitalisation.

Financing the Transition

With planning and forethought much of the €trillions required to finance the climate Transition could already have been put in place in building the new economic model required in response to COVID-19. Alongside those measures, there will be a need for the establishment of dedicated institutions—initially a European climate bank which then must be replicated at global level.

It cost €200 billion in today's money to put the first man on the moon. To return during the next decade will cost only €40 billion.[1] By contrast the sums of money to advance the Transition are huge and will cost the EU over €11 trillion just to comply with the Paris Agreement by 2030.[2] So when Ursula von der Leyen, the EU Commission President, said it was Europe's Man-on-the-Moon moment when launching the Green Deal, she was understating the challenge by several orders of magnitude.

[66] Schmidt P, The sustainable economy we need https://www.eesc.europa.eu/en/our-work/opinions-information-reports/opinions/sustainableeconomy-we-need-own-initiative-opinion. Accessed 3 June 2020.

The hard reality is that governments will be required to mobilise finance on a scale never seen before, even during a time of war or financial crisis. Achieving a climate-neutral economy within three decades will imply massive investments to renew our energy systems, our transport infrastructures and our production processes. A substantial part of this investment will have to target the building sector where greater efficiency will mean lower running costs and greater opportunities for profitable private investment. EU and national policy support is vital to reward the new portfolio of sustainable long-term investments.

The €4 trillion made available by the EU to counter the effects of the 2009 financial crisis will need to be substantially exceeded and current proposals, whilst welcome, are clearly just a beginning. Longterm, low-cost finance is needed for both climate mitigation (reducing the scale of the climate change) and climate adaptation (reducing risks deriving from climate change). The track record of European states in this area is not good. Finding a few billions to finance energy efficiency measures is already very hard at national or regional level, even in large European countries. But the main reason for the passivity of European governments is otherwise: they lack the necessary political will to raise and redirect the huge flows of public money required. There is no escaping the fact that future generations will have to pay for these resources, money needed to remedy previous mistakes. But this raises the important question of intergenerational equity. In many developed nations the wealth of those aged over 65, already substantial, is increasing faster than younger generations. In spite of this, intergenerational equity is not considered in policy-making, not least because older people tend to vote more regularly. Today's politicians lack the necessary political will to both spend a reasonable share of current budgets to mitigate future climate change and also to ensure that those largely responsible for the problem pay a proportionate share of the cost.

At present, the EC is arguing that the leading role in meeting the challenge will be taken by the European Investment Bank, Europe's "climate bank", which has developed an ambitious new climate strategy and energy lending policy.[3] However, the EIB also admits that their efforts will only 'unlock EUR 1 trillion of climate action and environmental sustainable investment in the decade to 2030' when at least ten times that amount is needed.

The EU must align all its public sector spending across Member States with the goal of achieving a climate friendly wellbeing economy. This does not concern only the main EU financing mechanisms, such as the Multi-Annual Financial Framework, the European Social Fund or the Common Agricultural Policy (CAP), but should apply to all budgets managed by the EU and lower institutional levels, down to the local one.

The political will of European leaders might be strengthened by the bold proposal made by Pierre Larrouturou and Jean Jouzel to create a European climate bank.[4] First formulated in early 2018, the proposal of the two Frenchmen was quickly endorsed by the European Economic and Social Committee (EESC) who presented their supportive opinion at COP24 in Katowice, before being backed in February 2019 by more than 600 political figures from 12 countries (including Spanish Prime Minister Pedro Sanchez). The proposal foresees the creation of two new tools to address global warming and social inequality: a green subsidiary of the European Investment Bank (EIB), to lend to member states, and an annual 100 billion fund to disburse grants across Europe, the Mediterranean and Africa. The bank's subsidiary would provide concessional loans for climate projects to each EU member state up to 2% of their GDP. That would amount to maximum 65 billion euro in the case of Germany and 45 billion euro for France. It is worth noting, that the school climate strikers of the "Rise up for Climate" movement in Belgium have decided to make the creation of the climate bank one of their central demands.

Their proposal is a practical response to the unambitious political thinking at European governments' level. In December 2019, the Dutch Supreme court upheld an historic legal order on the Dutch government to accelerate carbon emissions cuts, following a complaint by the pro-sustainability Urgenda Foundation,[67] together with 900 citizens. The Urgenda Climate Case was the first in the world in which citizens established that their government has a legal duty to prevent dangerous climate change. It took more than four years for the case to be handled by different levels of jurisdiction until the Supreme Court judgement. Referring to the severity and scope of the climate crisis, the judges of the different courts sent a clear signal to all governments: inaction on climate change is unjustifiable and new ways have to be found to finance truly ambitious climate plans at national level. The lack of financing has been one of the reasons why Mark Rutte's Liberal administration proposed a meagre 17% drop in greenhouse gas emissions by 2020 (vs. 1990), a target which the court judges deemed insufficient and asked should be raised to at least 25%. There could be around 900 similar court cases in the world, including in France, New Zealand, Norway, Uganda and UK.

The concrete proposals of Pierre Larrouturou and Jean Jouzel are to stop fossil fuel subsidies, combat fiscal evasion, intra-European fiscal competition and short-term speculation, and collect 5% of the corporates' profits to feed the annual €100 billion fund. Finding big sums of money is not the main difficulty. In 2008, the European Central Bank (ECB) put 1000 billion euro on the table to salvage the EU's financial stability. Since 2016, the ECB has created 2500 billion euro of liquidity to reinvigorate EU growth. In 1989, French President François Mitterand and Chancellor Helmut Köhl took 6 months to create the European Bank for reconstruction and development (EBRD), to finance the economic transition of the ex-USSR European countries. And in spring 2020, EU central governments have injected hundreds of billions of euro to prevent a massive economic fallout created by the COVID-19 crisis. On 27 May 2020, the EC proposed a 750 billion euro post-covid recovery plan geared towards the green and digital transitions, to ensure money used to repair the damaged economy will also drive the needed structural transformations in the long run. In a call made in June 2020 in support of the EGD as the best way to recover from COVID-19,[68] 700 personalities, including MEPs Pierre Larrouturou and Bas Eickout, Jean Jouzel and the young climate activist Anuna de Wever, have called on the ECB to invest

[67] https://www.urgenda.nl/en/home-en/. Accessed 31 May 2020.

[68] https://www.euractiv.com/section/climate-environment/opinion/european-call-3-solutions-for-climate-and-jobs/. Accessed June 4 2020.

annually at least 240 billion euro in climate and jobs. These billions should feed a Climate and Biodiversity Bank, which would provide interest-free loans to each Member State, up to 2% of its GDP each year for 30 years (i.e., €300 billion). The reasoning is that it is possible as the ECB has created €2600 billion since 2015 and that only 11% of these colossal amounts of money have gone into the real economy.

If all this could be done, the creation of a European climate bank or European climate and biodiversity bank should not be that complicated. The question remains—is the EIB's restated commitment to be Europe's climate bank and their pledge to stop funding most fossil fuel energy projects going to be sufficient or is a new, dedicated and focussed institution required?

Besides public money, there ought to be private money flowing into pro-climate projects. Financial investors, banks, pension funds and private companies must be mobilised. For this to happen, governments have to set the right market and regulatory incentives. The recourse to public-private partnerships must be streamlined in the climate protection sector. The potential is huge: according to Morningstar, the sustainable funds in Europe reached 668 billion euro at the end of 2019, growing by 56% versus 2018, and compared with a 18% growth of the European funds as a whole.[69]

As part of its Action Plan on financing sustainable growth,[70] the EU has set improved disclosure requirements on how institutional investors integrate environmental, social and governance (ESG) factors in their risk processes. A growing number of investors do not limit themselves to foster green business. They lend their money to companies embracing a larger set of sustainable development goals (as part of the 2030 SD Agenda). For instance, ESG factors can encourage investors to prefer companies creating decently paid jobs of quality.

Major investment funds and banks are gradually moving on disinvesting in fossil fuels and investing in long-term sustainability. BlackRock, which manages assets worth 6.9 trillion euro including big holdings in major oil producers such as BP, Shell and ExxonMobil, announced in January 2020 that it would put sustainability at the heart of its investment decisions, by using ESG factors for the constitution of their portfolio. It would no longer invest in thermal coal. There remains its huge portfolio in oil and gas but this is an indication of a new direction.

When talking of green financing by the private sector, there inevitably arises the issue of greenwashing, which has enabled large groups to cast an environmentally sustainable shadow on their marketing profile without touching upon their core business. Investors find it increasingly difficult to separate the wheat from the chaff when a growing number of firms pretend to be environmentally virtuous. There is no one size fits all definition of a "green fund". In the beginning of the 2000s, BP succeeded

[69]https://www.morningstar.co.uk/uk/news/199190/record-shattering-year-for-sustainable-investments.aspx. Accessed 31 May 2020.

[70]https://ec.europa.eu/info/publications/180524-proposal-sustainable-finance_en. Accessed 31 May 2020.

in providing itself a greener image whereas their investment in renewable energy was pathetically low. The oil spill in the Gulf of Mexico in 2010 forced reconsideration of these credentials.

To counter greenwashing and foster truly sustainable investments, the EU has established a unified classification system or taxonomy.[71] It took almost two years for the EU decision makers to agree upon a compromised proposal after having fought over the recognition of gas and nuclear as "transition" sources of energy. Bas Eickhout, the rapporteur on the file at the European Parliament, Vice-Chair of the Group of the Greens/European Free Alliance said: "Finally we now have an agreement on taxonomy that will shift financial flows towards more sustainable economic activities. It shows the EU and the rest of the world the direction where investments should be made. Financial products will need to prove their sustainability under strict EU criteria, it will encourage the markets to make sustainable investments the mainstream."[72,73] The compromise reached is that gas and nuclear can under no circumstances be included in the category of so-called 'pure green' investments, but this should not prevent those EU countries who wish to continue funding those industries.

The Transition will imply the decommissioning of fossil-fuel assets well before they have reached their full return on investment. Investors will require clear governmental signals and support on the regulation-driven depreciation of their fossil-fuel assets. Shocks to investors' expectations about the stringency and consequences of climate policies must be avoided as they constitute a risk for financial markets, which themselves are essential to operate a cost efficient Transition.

The role of climate action' ambassadors in helping funds to be raised should not be underestimated. Social media has proven to be an efficient vector for diffusing appeals for donations from web influencers. The number of funding appeals in this way is steeply increasing. Launched at the end of October 2019, the project #TeamTrees[74] aiming to collect $20 million to plant 20 million trees had received over $22 million by the end of May 2020. The project rests on the shoulders of new internet stars who have risen to the challenge of raising money through their young online followers. Typically such operations capture funds very quickly, and definitely more quickly than via the traditional means operated by associations and NGOs (typically postal mailing).

It is essential to improve the ability of ordinary people to invest their savings responsibly in ways that benefit the wider environment. Their confidence in the financial system has been largely eroded by the 2008 financial crisis and its dramatic consequences. This partly explains why fundraising via social media is becoming more popular. Young people are only a minority of the donors but they tend to be more

---

[71] https://ec.europa.eu/info/publications/180524-proposal-sustainable-finance_en. Accessed 31 May 2020.

[72] NASA Administrator Jim Bridenstine June 2019.

[73] https://www.greens-efa.eu/en/article/press/deal-reached-on-taxonomy-will-support-sustainable-investments/?pdf=32. Accessed 31 May 2020.

[74] https://teamtrees.org/. Accessed 31 May 2020.

generous than their seniors, as a proportion of their income.[75] Nevertheless, there is a major effort needed to stimulate response from those in later life. In the major European economies, this group, those who are 60+, control 70–80% of national wealth. Governments will need to find creative ways to incentivise those who are nearing the end of their life to give, save and invest in a sustainable future.

Last but not least, large fundraising operations, in the form of "climaction", have to be launched by governments with the objective to not only collect money for the climate but to also communicate widely to the public. In May 2020, the EC held a Coronavirus Global Response pledging event that registered €9.5 billion in pledges from various donors worldwide.[76,77,78,79] The works of the fundraisers and charities have to be facilitated by new government incentives.

If, at a minimum, the EU were to meet the '€11 trillion by 2030' challenge which the Paris Agreement implies, it would still be less than 5% of projected GDP for that period and would be a modest price to pay to set a standard of global leadership which could influence other major greenhouse gas contributing nations.

Acting Effectively at the Local Level

Climate change is a global threat that concerns all countries hence the international political process gradually moving forward through the COP meetings of the UNFCCC. There is a necessary logic in discussing the risks and the remedies at the international level as developing concerted, harmonised or even integrated approaches cannot but improve the overall efficiency of global action. Furthermore, the efforts of the international community to promote peace and multilateralism and to prevent commercial tensions have to be accentuated to safeguard its capacity to carry out joint projects.

There is however a reverse, equally strong justification to operationalise climate actions at the local level. One cannot ask people to be more engaged and participate in public decisions if the decisions are taken remotely at a stratospheric level. One cannot find the right way to combine and optimise policies if this integrated approach is decided far away from the implementation level. One cannot have the chance to tap into the vast pool of business creativity if again there is no proximity between the needs of communities and the possible solutions developed by enterprises. As

---

[75] Landscape review on EU Action on Energy and Climate Change, European Court of Auditors, 2017.

[76] Magazine Telerama, 4 December 2019, p. 22–28 (in French).

[77] As registered on 26 May 2020.

[78] https://www.eib.org/en/press/all/2019-313-eu-bank-launches-ambitious-new-climate-strategy-and-energy-lending-policy. Accessed 31 May 2020.

[79] https://www.pacte-climat.eu/en/. Accessed 31 May 2020.

highlighted by the Swiss developer of the solar plane "Solar Impulse", Bertrand Piccard, the deployment of cleantechs is increasingly decided at the local level.[80]

The Paris Agreement considers it essential to complement governmental climate action by non-state and subnational action, namely by businesses, civil society organisations, communities, cities and regions. Prominent international platforms, such as the Non-State Actor Zone for Climate Action (NAZCA)[81] by the Secretariat of the UNFCCC and the Climate Initiatives Platform (administered by the UNEP/DTU Partnership)[82] give some visibility to such action.

Climate Action Tracker[83] has reported a 2018 analysis of the U.S. GHG emissions that show the growing prominence of recorded and quantified commitments from sub-national and non-state actors in the US.[84] If all these commitments were fully implemented, they could result in emissions reductions of 17–24% below 2005 levels in 2025 and thus bring the country within striking distance of its NDC target. One of the main drivers of the U.S. states' and cities' commitment to reduce emissions has been the We Are Still In coalition[85] that was created in June 2017 to opposed to the US withdrawal from the Paris Agreement. On top of cities and states, the coalition gathers businesses, universities, healthcare organizations, and faith group, representing a constituency of more than half of all Americans, and taken together, they represent $6.2 trillion, a bigger economy than any nation other than the U.S. or China.

A 2018 study by the German Development Institute[86] showed that EU-led non-state action represents the majority (54%) of cooperative initiatives registered with NAZCA platform. However, implementation is not evenly distributed across Europe with actions in Central and Eastern Europe largely under-represented. There are many EU-based initiatives that are yet unrecorded. European-led action seem to perform rather well compared to the global average with a majority of them achieving a 'high output performance'. Most EU-led non-state climate action focus on mitigation but they do not set clear and quantifiable targets, complicating assessments of mitigation impacts. The main perceived challenges to greater and more effective climate action are the lack of access to finance, the lack of expertise and staff and the lack of recognition by public authorities. The strongest stimuli factors of non-state action are, by order of importance, external financial support, policy and legislative support, and recognition and credible communication of existing action.

The national climate plans must have a declination at the regional and city levels. Local climate protection plans, assorted with local climate targets, should be developed on the basis of the specific assets and weaknesses of the considered region. The experience is that where there are high levels of engagement and ownership by

[80]Interview in the La Libre Belgique newspaper, 11 December 2019, 22–23 p.

[81]https://climateaction.unfccc.int/. Accessed 3 June 2020.

[82]http://climateinitiativesplatform.org/index.php/Welcome. Accessed 3 June 2020.

[83]https://climateactiontracker.org/countries/usa/current-policy-projections/ 3 June 2020.

[84]America's Pledge, 2018; Data Driven Yale, NewClimate Institute, & PBL, 2018.

[85]https://www.wearestillin.com/ 3 June 2020.

[86]Chan and Bencini (2018).

local communities then there is greater acceptance of policy measures. Conversely, where policies are imposed from outside (national or EU level) and do not leave adequate scope to design local responses, or where projects are externally conceived and mostly owned by external actors (with few local benefits), then levels of local resistance can become a major barrier. In such circumstances, cultural sensitivities significantly condition the acceptance of the proposed policies within their local environment.

In short, the UN process must be paralleled by a multitude of regional processes that are framed by local authorities on the basis of local strengths and weaknesses. All governance levels, and not only the local one, have a responsibility in facilitating the emergence of ambitious climate plans at the grassroots level. These local plans have to be conceived with the same requirements that are applied to the national plans: i.e. a long term and integrated vision informed by the contributions of citizens, civil society and private companies. In the EU, national and local plans (e.g. under the Covenant of Mayors) should be aligned so that all governance levels pull in the same direction and their respective actions support each other.

The role of the lower institutional levels of governance, i.e. the regional and local authorities, is therefore fundamental. Europe will not reach its climate neutrality objective by 2050 as set by the European Green Deal without the involvement of cities, regions and local authorities. Although they are not signatories of the Paris Agreement, contrary to central governments, a growing number of European cities, regions and local authorities have taken the responsibility to contribute to the collective efforts to reduce carbon emissions. Some are even in advance of the national climate efforts by having decided to aim at climate neutrality at an earlier date than 2050. Helsinki has for example pledged to be entirely carbon-neutral by 2035, one of the world's most ambitious city climate targets.

The EU law obliges Member States to establish permanent multi-level dialogues that involve local and regional authorities in the definition of their National Energy and Climate Plans. According to the European Committee of the Regions (CoR), there has been too little attempt to establish a proper multilevel governance at Member States' level.[87] Yet, if the EU is to successfully make the Transition happen, it is critical to permanently involve local and regional authorities in developing the national climate and energy actions plans. The CoR also proposed to create a "European climate neutrality observatory", aimed at mapping and monitoring the vulnerabilities of the different territories to the Transition.

The involvement of sub-national authorities is not sufficient to stimulate bottom-up action. In line with the action point 3, civil society organisations have also a key role to play in mobilising citizens at local level and to ensuring engagement and participation of less informed and more vulnerable groups. Local and regional authorities should create the enabling conditions but civil society organisations are needed for awareness raising, capacity building, empowerment, etc. for citizens and communities to respond and benefit from these enabling conditions. The reality is that the most innovative climate-friendly practices are mostly decided

[87] A Clean Planet for all, European Committee of the Regions, opinion CDR 5736/2018.

and implemented at the regional level; this is notably the case of organic farming, short production/consumption circuits or new experiences of sustainable community living, the latter in an attempt to apply the bioregional vision of U.S. author Kirkpatrick Sale.[88] This is why the relationship between the local (local authorities) and sub-local (communities, neighborhoods) levels is so important.

Historically, the affirmative role of sub-national authorities to reduce climate change emerged a decade ago[89] with the establishment of the EU Covenant of Mayors for Climate & Energy.[90] Launched in 2008, it aimed to gather local governments voluntarily committed to achieving and exceeding the EU climate and energy targets. Since then the EU Covenant of Mayors for Climate & Energy has expanded internationally and the initiative now gathers more than 10,000 local and regional authorities across 60 countries, representing almost 320 million citizens. Now expanded to the Global Covenant of Mayors, this worldwide movement has developed a wealth of positive experience with cities actually reducing carbon emissions through citizens involvement and innovative solutions, and it sets a benchmark in terms of providing technical and methodological support through their local offices.

There are many regional and global movements that promote climate-friendly initiatives at the local or sub-local scale, either bringing together big cities, or bringing a mix of cities, stakeholders and citizens. C40 is a network of the world's megacities committed to addressing climate change by supporting cities to collaborate effectively. The Compact of States and Regions brings together leading states, provinces and regions in ambitious efforts to respond to the Paris Agreement. The Under2 Coalition is a subnational government climate action network, led by state and regional governments and bringing together over 200 signatories from six continents and 43 countries committing to reduce their greenhouse gas emissions by 80–95% compared to 1990 levels. The United Nations Global Compact is a voluntary initiative based on CEO commitments to implement universal sustainability principles and to undertake partnerships in support of UN goals. The Transition Movement comprises local communities taking concrete actions to address climate and other sustainability challenges. ECOLISE is a coalition of national and international networks of community-led initiatives on sustainability and climate change, which includes the Transition movement, as well as networks and organisations that support a community-led transition to a resilient Europe.

To accelerate the sharing and replication of good practices, more collaborative networks of this kind are needed and governments must support their creation and operation through various measures. Priority should be given to networks such as Transition or ECOLISE promoting multi-stakeholder approaches and encouraging the empowerment of citizens and communities. The different levels of governance can create the enabling conditions but this is not enough. Citizens and communities still have to be steered towards climate action. Multi-stakeholder networks tend to take a

[88] Sale (1985).

[89] Agenda 21 already supported similar action since 1992.

[90] https://www.covenantofmayors.eu/en/. Accessed 3 June 2020.

more holistic approach, not just focusing on carbon emissions but on 'transitioning' away from a system that is inherently damaging to the environment and to society.

In the same way as for the central governments, sub-national authorities must become quickly aware of the multiple benefits that can be derived from conceiving and implementing decisions in common, among various stakeholders. The multi-stakeholder networks are best placed to tap into the very different territorial dynamics in order to propose specific solutions for the local communities. By bringing all the key actors and citizens together, there is higher chance to mobilise around common objectives and create local dynamics that benefit from the required human and financial resources. The role of these multi-stakeholder networks is therefore key for quickly and effectively diffusing positive climate action within and between local communities.

Last but not least, there must be a close connection between the local multi-stakeholder communities and the central governments, with the aim to align bottom-up and top-down approaches and action. The three levels of governance from city level to national level via the intermediate regional level must boost their cooperation in order to ensure more consistency and efficiency of their different climate plans. Central governments are called to set up efficient multi-level and multi-stakeholder governance mechanisms to amplify climate change mitigation and adaptation efforts.

Paving the Way to a Reform of Economic Policies

The period between the fall of the Berlin wall in 1989 and the financial crisis in 2008 could possibly remain in history as the last time when it was possible to argue that the free-market economy was still the solution to bring about a stable economic order. Nowadays, a growing number of people and organisations think that boosting multilateral trade, rising labour productivity and keeping wages under control cannot help to resolve day-to-day problems faced by the European population, not to mention the more critical challenges such as climate change. If such a consensus no longer exists, it is principally because the arsenal of measures employed since 2008 has not significantly improved the economic, social and environmental situation; it is as if the recipes of the past have become ineffective. The liberal economy, with its narrow focus on growth and competitiveness no longer enhances the well-being of people, that is the hard fact of today.

"Our" economic model, the one through which countries pursue economic growth as their top policy objective, is no longer self-evident. It has brought enormous benefits to people in the past, especially in high-income countries. But today, economic growth is slipping away, after having shown signs of decline already before the 2008–2009 financial and economic crisis. Measures to counter faltering labour productivity growth have not proved effective. The sluggish economic expansion in developed countries, with its enervating impact on public finance is leading to the reverse effect that it had in the past: an increase of income and wealth inequalities, unsustainable levels of debt and financial instability, combined with a shrinking of the social security net.

Confronted with a multitude of challenges, principally of a social and environmental nature, the dominant economic model based on the GDP has lost the capacity to adapt itself and bring practical solutions whenever they are needed. As stressed by Tim Jackson, Director of the Centre for the Understanding of Sustainable Prosperity (CUSP) and Professor at the University of Surrey in the UK, it is our obsession to prop up an ailing capitalism with all sorts of pro-growth tricks (low interest rates, cheap money, inward investment, government stimulus, tax havens, fiscal austerity, and customs partnerships) that has diverted our attention from more essential priorities such as climate change, resource bottlenecks and social stability. The technological toolbox is likely to continue to offer further ways to improve labour productivity. Artificial intelligence is following automation, both winning massive backing from investors and policy-makers alike. The cost is a continuing decline in less skilled jobs and this must challenge assumptions about the stability of our economic system.

As emphasised before, this book aims to counter the easy argument that "there is no alternative",[91] the (in)famous TINA. Finding an alternative to today's dominant economic model based on the GDP fetish should become the next big question being asked by the international community. Many green activists consider that, to be credible, environmental movements have necessarily to be anti-capitalistic. In their view, the size of the required transformation is such that policies reliant on a framework of capitalism have to be scrapped. The last 50 years, with their repeated environmental catastrophes (Seveso in 1976, Bhopal in 1984, Exxon Valdez oil spill in 1989, Fukushima in 2011 etc.), have shown the fundamental inability of capitalism to take into consideration the environmental issue, other than superficially. On the other hand, there are also numerous examples of similar failures by centrally planned economies, Chernobyl being a prime example.

Citizens are increasingly being informed and confronted with the inefficiency and weaknesses of today's main economic model and yet know that previous systems have also failed in this respect. Mentalities are evolving fast. For a still small but growing share of the population, having the business sector adapting their strategies at the margin would not be sufficient. An increasingly loud call to proceed with an in-depth reform of economic models is being heard. The civic pressure should make the demand for a new economic model one of the key measures to meet climate neutrality by the mid-century. However, although there is a relative consensus on the dysfunction of today's system, there is hardly any convergence of views on the solutions to reform it.

Symptomatic of this lack of consensus is the way the degrowth concept is being diversely understood. Though having earlier origins, the concept of degrowth has entered the public sphere in 1972 with the publication of the report of the Club

[91] TINA (There is no alternative) was often used by Margaret Thatcher, former UK Prime Minister, to reject her opponents' ideas.

of Rome, called The Limits to Growth.[92] The report, also known as the Meadows Report, is not per se a degrowth report as it rather advises a zero growth. However by conveying the message of the downscaling of production and consumption to prevent overconsumption and its detrimental long term environmental and social impacts, the report has often been associated with the promotion of degrowth.

Applied dogmatically, degrowth could rapidly lead to devastating results. Producing and consuming less, though justifiable from a purely environmental point of view, would have the immediate consequence of leading to factory closures, spreading unemployment, public money deficits, social benefits steep reduction etc. It is therefore quite telling about today's society that The Limits to Growth report is still so popular. More than 30 million copies of the book have been sold in 30 languages. And a 40-year update has been published in 2012.[93] Its scientific and political impact has been and is still considerable today. The reason might be that it identified a fundamental long-term weakness in the accepted economic model.

The covid crisis has dramatically highlighted the vulnerability of an economic system built on evergrowing GDP. Up to 10% of GDP could be lost in the first year of the crisis[94] with uncounted millions of people losing jobs and homes. The resurgence of interest in degrowth since the previous devastating financial and economic crisis in 2008, has therefore to focus on the adjustments needed to soften the social and environmental impact of reduced or zero growth—not in a crisis situation but as part of a new economic model. One potential positive outcome from the covid crisis could be to stimulate further exploration of alternative economic models, including variants of degrowth.

The word degrowth does not enthuse or inspire people because it is negatively connoted with regression and decline. To make it more politically acceptable, many now prefer to speak about post-growth, to reflect that growth is an inherent feature of human societies but that growth must be more of a different nature. The most tangible sign of a greater openness towards degrowth has come from Pope Francis in 2015 in his Laudato Si second encyclical.[95] Focusing on the care needed for our common home, Pope Francis recognised that "time has come to accept a certain degrowth". The words degrowth and postgrowth are increasingly employed in the media but they continue to be very cautiously used by the politicians who fear being branded as extreme ideologists.

Whatever words or concepts are used, it is plainly the responsibility of governments to explore the more plausible options to renovate the current economic model. Plenty of economists have fused their neoclassical economy framework with other disciplines assessing social and environmental impacts and in doing so have researched ways for improving today's economic policies. Many proposals of revised economic models have the inherent capacity to promote social and environmental

[92]Meadows et al. (1972).

[93]Randers (2012).

[94]Asian Development Bank forecast May 2020.

[95]Laudato Si—On Care for Our Common Home, Pope Francis, ed. Our Sunday Visitor, 15 July 2015, 176 p.

progress, besides economic performance. A taboo-free, pro-responsible-business approach is essential to ensure objectivity in analysing the many dimensions of a possible reform of economic policies.

There are many possible measures for strengthening our economy at times of pressing social and environmental challenges. The three that offer the greatest potential of transformation and require the greatest courage of governments are:

1. Replacing the GDP by a new environmental macroeconomics

Moving beyond GDP has been the subject of intense research since the end of the last century. Invented in the US in the 1930s, GDP is an incomplete, economy-centred indicator that cannot take due account of the full range of policy objectives that must be considered as key by governments (e.g. health, education, social justice, quality of the environment),. Official statistics centred on GDP do imperfectly match people's lived experience. For instance, the US economy is experiencing its longest period of high GDP since the end of the second world war (the covid fallout apart) and yet many people have not had the chance to feel that upside; there were around 38 million people living in poverty in the USA before March 2020.

The problems with using GDP as a leading indicator go beyond masking inequality. Polluting factories count in the GDP metric but there is no measure for a clean environment, quality of life or sustainability in general. Narrow GDP-based economics can present a distorted view of a country as GDP is morally indifferent to factors like pollution, welfare and human rights, financial speculation and certain types of crime. Mis-measuring our lives means losing the capacity to assess the real state of the society and to correct vital deficiencies. It could have very destructive consequences for democracy as the real needs of the people are simply not part of the big picture.

There have been many proposals for replacement of GDP by sets of multi-policy indicators but these have remained on the shelves of research departments and policy institutions. So far there has been almost no concrete attempt to proceed with the replacement of the GDP by such indicators at national or regional level. Moving from narrow economic goals to a varied ensemble of societal goals, is not an impossible task as shown by the very positive experience undertaken in Bhutan. The U.K. has been measuring national happiness since 2011.[96] This would imply imbuing the fundamental functions to the state with stronger well-being and sustainability perspectives.

An interesting example of a measure of economic and social welfare is the Genuine Progress Indicator (GPI).[97] Meant to replace or supplement GDP, GPI has been conceived to take fuller account of the wellbeing of a nation by incorporating environmental and social factors as modifiers on the traditional economic factors measured

[96]Personal wellbeing in the UK: Office of National Statistics.

[97]Lawn and Clarke (2006).

by GDP. For instance, GPI decreases in value when the poverty rate increases. Similarly, GPI will be equal to zero if the financial costs of poverty and pollution equal the financial gains in production of goods and services, all other factors being constant. It is "informally" used in the US and Canada but GDP remains the official development indicator in these two countries.

Professor Tim Jackson, has made very concrete proposals to replace GDP by a new environmental macroeconomics, notably in his book on "Prosperity without growth".[98] With Professor Peter Victor from York University, Canada, they have developed the conceptual basis for an economy in which stability no longer depends on relentless consumption growth. The new environmental macroeconomics that they call for is a clearly distinct proposal to the green growth one which, despite being more researched and mediatised, does not have the capacity to mitigate resource demand in absolute terms in a growing world. They provide very practical recommendations to transform the economy in ways that protect employment, facilitate social investment, reduce inequality, enhance quality of life and deliver both environmental and financial stability. Their proposal is consistent with the conclusions of the Commission on the Measurement of Economic Performance and Social Progress (CMEPSP), which was created by the French Government in 2008 and was chaired by Nobel Prize -winner, Joseph Stiglitz.

In an opinion adopted in January 2020, the European Economic and Social Committee (EESC) proposes a new vision of prosperity for people and planet embedded in the wellbeing economy. The EESC opinion entitled The Sustainable Economy We Need[99] recognises the need to support the fundamental changes that have already begun to emerge in the nature of enterprise, the organisation of work, the role of investment and the structure of the money system. Building the wellbeing economy must start by adopting a precautionary approach in which macroeconomic stability does not depend on GDP growth. It is therefore proposing the development of new indicators of economic performance and social progress beyond GDP, and the introduction of a wellbeing budget for the EU, drawing on the example of the New Zealand Living Standards Framework.[100] A review of the growth dependency of the EU Member States has to be carried out for designing a proper strategy to enable the EU economy to focus on sustainable and inclusive wellbeing. The reform of the GDP into a wider macroeconomics could have a booster effect on public engagement in meeting the climate challenge.

2. Regulating the market

One way for governments to move away from the vicious spiral of exhausting human and environmental resources is to change the macroeconomics and to base their long-term strategies on a wider set of goals centred on the well-being of people. This will

---

[98] Jackson (2016).

[99] Schmidt P, The sustainable economy we need https://www.eesc.europa.eu/en/our-work/opinions-information-reports/opinions/sustainableeconomy-we-need-own-initiative-opinion. Accessed 3 June 2020.

[100] https://treasury.govt.nz/information-and-services/nz-economy/higher-living-standards/our-living-standards-framework. Accessed 3 June 2020.

require that governments accept a responsibility to regulate whenever the market economy is leading to social, economic and environmental imbalances. The power to regulate has to be reinstalled as a more active tool of government.

Social and economic imbalances significantly reduce the efficacy of environmental policy measures. For all people in society to feel the same need to act in favour of the climate, wealth has be more equally shared. One of the main barriers to the environmental transformation of society is the persistence of important wealth' gaps in a majority of European countries, pointing to the rising inequalities created by free market economies. In Europe, the share of the national income that goes to the 10% richest people raised from 33 to 37% since 1980 whilst it moved from 34% to 47% in the US. As noted by Thomas Piketty in his book on Capital and Ideology,[101] at stake is the rebuilding of democratic deliberation in order to create new norms of social, educational, fiscal and climate justice. These new norms will have to be profoundly equitable to enable all actors in society to feel empowered and respected in their own rights and capacities, a precondition to anyone acting for the climate. Similarly to Thomas Piketty, a growing number of economists[102] call for the return of tax progressivity to be a pillar of these renovated norms.

In his book on People, Power and Profits,[103] Nobel prize Joseph Stiglitz validates this approach. His country, the USA, has become a society of annuity's exploitation with no consideration for workers, people and the environment. The industrial concentration has reached an historical high, especially in the digital sector (e.g. GAFAM[104]), and there seems to be no attempt to break it up through an equivalent of the anti-trust laws used in 1911 to split Standard Oil. This hyper-concentration gives a free hand to oligarchs to endlessly boost productivism and consumerism, exacerbating the dependency of workers and consumers. Oligarchies not only amass excessive economic power but also a political force which limits the governments' capacity to take positive actions for people and the environment. In other words, it is becoming increasingly difficult for governments to reform economic policies and regulation as the big multinationals and economic powers have become an inherent and influential part of today's political process.

In 1942, Franklin Delano Roosevelt proposed to the U.S. Congress to create a 100% marginal rate of imposition for incomes above 25,000 dollars (around 1 million dollar in current values) and the Congress finally agreed to fix it at … 93%. Part of the argument of Roosevelt is still valid today: concentration of wealth means concentration of power and capacity to influence the markets and impose an ideology. Today the big conglomerates are the fiercest partisans of the economic status quo and their capacity to adapt to the frequent bumps has no limit. The GAFAM will emerge stronger from the coronavirus crisis.

By itself, the market does not punish businesses that are harming society by exhausting the natural resources or pursuing cost reduction in an unrestrained manner.

---

[101] Piketty (2020).

[102] Saez and Sucman (2019).

[103] Stiglitz (2020)

[104] Google, Apple, Facebook, Amazon, Microsoft.

It is the plain and legitimate role of governments to counter the market when some companies are abusing the social and environmental forces of the society. Among the measures that Joseph Stiglitz calls upon to create a progressive capitalism is the dismantling of conglomerates. As a general matter, all organisations' abuses, that harm the society and the environment, must be more strictly controlled and punished through regulation.

Taking advantage of the new situation created by the covid crisis, new regulations are needed to combat growing inequalities, create an equal footing for all organisations and people to feel like-minded and uniformly enthused to do good for the environment and put the economy at the service of the society as a whole. In its opinion on The Sustainable Economy We Need,[105] the EESC asked the EU governments to carry out green fiscal reform to help align taxation, subsidies and pre-distributive policies with the goal of achieving a just transition to a wellbeing economy, in particular by enforcing existing legislation.

3. Promoting sustainable trading practices

The mitigation of international trade is another route to explore. It is certainly not consensual, depending on whether you find yourself in the group of winners or losers of the globalisation process. Until recently, Europe has largely benefited from the globalisation of its economy but this could change in the future. To renounce globalisation on the grounds of developing more local and sustainable cooperative approaches seems tempting but there is a long way to go to achieve it in practice. The covid crisis has exacerbated demands for relocalising national production but a proper assessment of such measures must be a prerequisite given the conflicting interests involved. European car manufacturers are eager to expand their market internationally whereas European farmers are increasingly opposed to trading with countries that are hyper-competitive in the agro-food sector thanks to their lower income or laxer environmental and sanitary regulations.

The promotion of international trade remains the official line of the European Commission, the executive body of the European Union (EU), which sees it as a way to promote multi-lateralism and the core EU values of democracy and the rule of law. The promotion of sustainability and of climate neutrality in particular, is however much more difficult to achieve in the Free Trade Agreements (FTAs) that the EU concludes with third countries as one frequent feature of FTAs involves long distance freight. The International Transport Forum (ITF), which is part of the OECD, estimates that international trade-related freight transport currently accounts for around 30% of all transport-related CO_2 emissions from fuel combustion, and more than 7% of global CO_2 emissions.[106] Furthermore, growth in international

[105] Schmidt P, The sustainable economy we need https://www.eesc.europa.eu/en/our-work/opinions-information-reports/opinions/sustainableeconomy-we-need-own-initiative-opinion. Accessed 3 June 2020.

[106] The carbon footprint of global trade—Tackling emissions from international freight transport, OECD, International Transport Forum, 2015, https://www.itf-oecd.org/sites/default/files/docs/cop-pdf-06.pdf. Accessed 3 June 2020.

trade has been characterised by more geographical fragmentation of international production processes, as the covid crisis acutely reflected. Supply chains have become longer and more complex, as logistics networks link more and more economic centres across oceans and continents.

In the majority of existing Free Trade Agreements (FTAs), there is no real consideration for sustainability. However, following the 2011 EU-Korea FTA, advisory groups are regularly set up in the EU and in the partner country or countries to advise on the implementation of the sustainable development chapters in EU trade agreements. Those chapters contain commitments to respect multilateral labour and environmental agreements and to ensure that labour and environment standards are not lowered by joint trading. The sustainable development chapters of the new trade agreements are prepared on the basis of ex-ante sustainable impact assessments (social/human rights, environment, economy etc.) that are carried out at the beginning of the trade negotiation.

Though these recent developments go in the right direction, there is much more to be done. In particular it is essential to carry out objective ex-post assessments of all FTAs, not only the ones concluded before 2011. When required, corrective measures to moderate excessive climate impacts must be taken.

A positive signal has come from Ursula von der Leyen as part of her proposal for a European Green Deal. If one takes seriously her wish that climate becomes an irrevocable pillar of EU FTAs, it is high time that EU governments have the courage to address the necessary reforms of their free trade policy.

For the above substantial reforms to be effectively passed at central government level, the left-right chasm will have to be constructively bridged to give a chance to decision-makers to tap into the full range of options at their disposal. The business sector will in particular have the challenge of assessing whether the infinite pursuit of quantitative expansion is the best way to respond to consumers who increasingly value quality in its widest dimension (product quality, decent labour conditions and decent salary, environmental quality, etc.).

How can business profitability be maintained in a society consuming fewer natural resources? How could a European reform of its economic policies be matched with the international context? Are market-based instruments, such as the EU carbon trading system,[107] enough to drive the steeper international reduction of carbon emissions in the next three decades? Should international trade be moderated on the grounds of preserving the climate and enabling local communities to become revitalised by recreating local production facilities. There are plenty of essential questions for governments to discuss, openly and without taboo. Tomorrow's macroeconomics will have to be based on a wider set of goals and more justice throughout, while keeping essential ingredients of today's liberal system such as enabling a certain level of competition to stimulate the emergence of new initiatives and technologies. Europe is not married to GDP as a growth indicator, 10 years from now it may look totally anachronistic.

[107] Called Emission Trading System (ETS), in place since 2005, https://ec.europa.eu/clima/policies/ets_en. Accessed 3 June 2020.

Making Sustainable Consumption Mainstream and Popular

Similarly to macroeconomics reform, the promotion of sustainable consumption is a specific policy that, by its complexity and political sensitivity, requires particular backing and processing from governments. Both need integration in the manifesto on climate governance.

There are many ways to trigger more sustainable consumption patterns and the intention herein is not to be exhaustive but to highlight those approaches which deserve a special and determined action by policy-makers and politicians.

Within European society, there is slowly emerging an increasing appetite to consume differently. People are becoming environmentally and socially conscious and a growing minority are testing new ways of consumption that better reflect their values and lifestyle. The specific demands are many, from organic to local via fair trade, but they tend to coalesce into a wider, sustainability-centred demand, with potential gains towards carbon emission reduction.

What people and organisations consume is dependent on the market which itself is largely conditioned by two factors: trends in consumer behaviour and the regulatory context in various policy domains. Technologies, as well as technical norms and standards, are also largely influencing the way the market operates. The covid crisis, with its major market disruption in different sectors of activity, has plainly demonstrated that the collective power of consumers is immense. To give the sustainable consumption movement, principally urban and still in the minority, a growth perspective will require another, more determined level of intervention from governments and a similar engagement of the business sector to bring to the mainstream sustainable products and services.

Unsustainable consumption is largely fuelled by the market and its tendency to drive costs down which has devastating impacts on the working conditions along the supply chain. Food can be produced by farmers who are earning much less than the living wage, as if food would have no cultural and societal value, like a mere commodity product. Moving away from this noxious trend requires a dramatic change in our economic models (action point 9). Neither is it true that the responsibility for sustainable consumption only rests with consumers, as if producers, distributors and retailers are exempt from accountability. The large retailers, in particular, are forcing prices downwards which enhances a products' affordability but is detrimental to sustainability in general (low wages, over- production, lower environmental standards). The responsibility for sustainable consumption must be equally shared throughout the supply chain from the raw material producers to the final consumers.

Though the EU has put in place one of the most complete sets of legislation to promote sustainable consumption (besides sustainable production), there remains a substantial task ahead if the objective is to quickly dis-incentivise the consumption of climate unfriendly products and services. Significant, but limited, gains by a minority of consumers are largely offset by the hyper-consumerism of a large majority of the population. Believing that everyone will become sustainability conscious is illusory so penalising t excessive consumption should become a declared objective of decision makers. Accordingly the time has come to critically assess the efficiency of current policies. Loopholes and low outcomes have to be corrected by a revised mix of carrot and stick measures. New, restrictive limits have to be set and vigorously enforced for the most unsustainable products and services while positive incentives have to stimulate the consumption of their sustainable equivalent.

The number of measures or pieces of legislation that could drive the transformation is immense, but it is a worthy regulatory investment. The existing legislation has principally been based on carrot measures, steered to some extent by technological innovation. An interesting example of this approach is the set of EU directives that aim at the reduction of energy consumption of energy-intensive products. These policies are in place since 1992.[108] They aimed in particular at enabling customers to make informed choices based on the energy consumption labelling of energy-related products. Providing information on the most efficient and sustainable energy-related products offers the potential for significant energy saving and energy bill reduction. There were however a series of issues that were not correctly planned and led to a lower outcome of the legislation.[109] The fact that the most energy efficient devices are disproportionately more expensive than efficient ones and the issue of planned obsolescence blurring even more the benchmarks, have contributed to a lower yield of this flagship EU policy. After some years of implementation of the EU energy labelling regulations,[110] a majority of products belonged to the top classes of the energy label (the 'A' class for the most energy efficient products), making it more difficult for consumers to make informed choices in favour of more efficient products. Furthermore, there was a fair amount of non-compliance at EU countries' level, caused by weak enforcement by national market surveillance authorities, leading to a loss of about 10% of envisaged energy savings.

Another crucial problem that has arisen relates to the so-called Jevons paradox[111] which leads to more use of a resource when the efficiency with which the latter is used is increased thanks to technological innovation. In his book on "The coal question",[112] William Stanley Jevons, an English economist, observed that, despite steam machines being more energy efficient in 1865, much more coal was consumed compared to 1800. The same phenomenon still occurs today, with a steady increase of usage of equipment whose energy consumption has been progressively reduced over time. This is particularly marked when the source of energy is electricity, whose cost is rather abstract for consumers (the invisible cost of battery recharging). There are numerous examples of increased number and size of refrigerators and domestic appliances observed over time as they become more efficient. The Jevons paradox has

---

[108]Council Directive 92/75/EEC on the indication by labelling and standard product information of the consumption of energy and other resources by household appliances, 22 September 1992.

[109]Commission staff working document, Impact assessment, Accompanying the document "Proposal for a Regulation of the European Parliament and of the Council setting a framework for energy efficiency labelling and repealing Directive 2010/30/EU, SWD(2015) 139 final, 15/07/2015.

[110]15 groups of products require an energy label.

[111]Bauer and Papp (2009).

[112]Jevons (1865).

been used as an argument by some that it is counter-productive to pursue sustainability by increasing energy efficiency. Often the energy savings forecast from additional home insulation are lower than planned as occupiers increase their level of comfort, a not unreasonable response by disadvantaged or vulnerable groups. Even if the EC began to consider the rebound effects caused by technological innovation,[113] they have generally not been taken into account in policy-making.[114] If the rebound effect is usually ignored, it is due to the lack of knowledge about the intensity of its potential indirect and economy-wide effects.

Applying disincentives through legislative measures is therefore the required paradigm change to be realised by governments. Hyper-consumerism should be heavily reduced through, for example, restrictions on harmful advertising (particularly to children), the regulation of harmful products, the promotion of ethical and sustainable consumption choices, the encouragement of community endeavours and the protection of public goods.[115] Taxation is indisputably the best means to steer consumption towards sustainable options. To increase the price of "dirty" options, there is a simple and well-known recipe: the internalisation into the pricing structure of the environmental costs induced by the production of goods or services. The internalisation of these "external" costs has been the object of hundreds of studies that have unanimously recognised its efficiency. It has been on the political agenda of the EU since 1995,[116] especially in the transport and environmental policy domains. The EC stressed its value in several policy papers, and in particular in a Communication on a strategy for the internalisation of external costs[117] where it suggests to recourse to a range of economic instruments such as taxation, tolls and CO_2 emissions trading.

Yet governments have largely ignored recommendations from the EU executive and from science in general. They have continued to let the market decide by itself, an approach which must now be scrapped. The internalisation of external costs must also consider the social costs created by poor working conditions or low wages. The EU legislation on the energy labelling of products is typical of the lack of courage of politicians: the equipment items that are the most energy efficient should also be more widely affordable to less wealthy households. The pricing measure has to be designed in such a way that it does not discriminate against the poor, to be socially

[113]Maxwell et al. (2011).

[114]With the exception of one UK policy to improve the thermal insulation of households (providing for the possibility that some of the potential benefits of the measure will result in higher internal temperatures, rather than reducing energy consumption; see DEFRA (2007), Consultation document: energy, cost and carbon savings for the draft EEC 2008–11 illustrative mix, Department of Environment, Food and Rural Affairs, London.

[115]Schmidt P, The sustainable economy we need https://www.eesc.europa.eu/en/our-work/opinions-information-reports/opinions/sustainableeconomy-we-need-own-initiative-opinion. Accessed 3 June 2020.

[116]The Green Paper on Towards fair and efficient pricing in transport—Policy options for internalising the external costs of transport in the European Union, European Commission, 1995, COM(1995)691.

[117]Communication on a strategy for the internalisation of external costs, COM (2008) 435 final, 8/07/2008.

sustainable. This implies that there always are cheap options available on the market not only for essential products and services but also for sustainable ones.

In the same vein, taxation should relate to level of use and not exclusively to ownership, as it is the case in a large majority of products. The excise duties on road fossil fuel is one of the rare examples of taxation that offers the potential to dissuade overuse of internal combustion engines. An increasing number of car insurance companies propose contracts whose price is modulated as a function of the number of kilometres driven annually. For the most unsustainable products, there should be a double disincentive: high acquisition taxes and high utilisation taxes.

Governments must complement their pricing policies with additional regulatory measures that can further dissuade an excessive or improper utilisation of certain goods and services. The installation of household electricity smart meters is one of those potentially efficient measures that has already been implemented in several European countries. Under a 2009 law,[118] EU member states are required to deploy smart meters based on a cost-benefit analysis with the goal to reach 80% of connection points by 2020.[119] There is however no plain consensus on the value of rolling out smart meters in European homes and a country like Germany has challenged the EU decision until recently. Whilst leaving some more time for Member States to achieve deployment, the latest EU legislation[120] goes further in that direction with stricter provisions on regular cost-benefit analyses and recommendations to electricity companies to provide energy management services. In addition, it provides that final customers are entitled to have a smart meter installed on request, where smart meters are not systematically deployed, and that the benefits of smart meters for consumers will be regularly monitored.

The argument used by the promoters of the technology is that enabling consumers to know in detail their different appliances' consumption and peak periods would naturally lead them to adjust their pattern of consumption. In practice, this is far from the case as there is very little adaptation made by households as a result of knowing more precisely how and when they consume, because of the complexity that it entails. To really help households to reduce their consumption, the adaptation should be automatic, with smart meters relying on time-of-use tariffs combined with home/appliances automation within a complete chain of smartness and services. If high consuming equipment such as fridges and freezers would themselves be smart, they could automatically be switched off during peak hours, delivering net advantages to the consumer.

The promotion of a collaborative or sharing economy is another trendy but realistic option that has been gaining much traction within society for the last decade or so. People are progressively realising that owning equipment that is rarely used is a costly non-sense. Leasing or sharing with neighbours, friends and even strangers are

[118]Third Energy Package.

[119]A study from December 2019 on the deployment of smart meters in the EU found that close to 225 million smart meters for electricity and 51 million for gas will be rolled out in the EU by 2024. This represents a potential investment of €47 billion.

[120]Directive (EU) 2019/944 of the European Parliament and of the Council).

spreading practices. But much more could be achieved if governments would adopt regulatory measures to facilitate operation of the sustainable business models active in the collaborative and sharing economy and reward users.[121] Such new business models must be developed while prohibiting substandard working conditions.

To enable low-income households to acquire the most sustainable goods (when the lower taxation of these goods is not sufficient to make them affordable), the system of third party investor has to be fostered, again with the help of governments and regulation. This is in particular needed for high CAPEX equipment such as energy efficiency and renewable energy generation measures (solar panels, heat pumps etc.) for housing.

Introducing individual carbon budgets is another interesting idea that is worth considering. Combined with a ranking of products based on their environmental impact/GHG content, such a system could help consumers to better identify the purchases that are carbon intensive, helping them to make alternative choices that could reduce their own carbon footprint—and keep it within their budget.[122] Combined with a bottom-up communication and persuasion campaign (action point 2), inspired by influential civil society figures, it could step by step lead to people reflecting twice before clicking on a purchase order. There are many websites offering tips to reflect on ones' essential needs and to inspire carbon footprint reduction in a way which creates positive psychological feedback.[123]

Enforcement of policies by national and regional governments is also a must to achieve the expected environmental outcomes. Many impact assessment studies carried out by the EC showed that there are recurrent problems related to the implementation of EU policies caused by a lack of enforcement by authorities. This has been notably revealed in the Dieselgate scandal, which had as direct consequences to lead to reduced energy and environmental savings and to consumers being cheated because their energy bills were not reduced by as much as was announced by the car manufacturers. Governments have to stay firm in the application of legislation, a *sine qua non* condition to retain trust in the complex relationship between the citizen and the legislator.

Last but not least, the business sector has a major role to play in creating new demands for sustainable products and services. Lack of positive engagement by business will simply nullify any communication campaign (action point 2) or progressive reform of the economic regulatory framework (action points 1 and 9). The business sector has to anticipate fashionable trends and upcoming legislative development.

Not only does it have to place the protection of climate as a central goal of its marketing, in order to "create shared value", but it needs to contribute to the creation of a new macroeconomic approach. The design of products and services that are fashionable, durable and carbon frugal must become their number one goal even though it may involve radical change. Underlying this is the issue of human desires and the way

---

[121] Such as through dedicated lanes or authorising to use bus lanes for car-sharing.

[122] Whose level could be fixed by governments on the basis of the national GHG reduction targets.

[123] https://green.harvard.edu/tools-resources/poster/top-5-steps-reduce-your-energy-consumption. Accessed 3 June 2020.

in which they tend to be manifested in consumption. The business sector must trigger new forms of desire for sustainable products and services. Consumers' demands must be subject to redirection in order to limit over-consumption of resources and the role of the business sector in this transformation is essential.

Conclusions

This book has shown that if governments can take a lead in the changes needed to bring about a new economic system, and if they are supported by the public and implemented by business, organisations and local communities, there is a real possibility of averting the looming climate catastrophe. This will be achieved:

- by reinvigorating public debate and re-engaging citizens in the life of countries and regions
- by reorienting capitalism towards production, innovation and the Transition
- by investing massively into citizenship, awareness raising, science education, technologies
- by rebuilding an alliance of democracies to promote cooperative action
- by reconnecting people with the defence and promotion of essential values
- by rewarding those people seeking to live more sustainable lives
- by dissuading hyper-consumerism
- by reconciling diversity of cultures, social rights, and the protection of the environment
- by placing first the wellbeing and dignity of humankind.

The journey towards a climate friendly society begins with a shared vision, anchored on a narrative of hope and confidence, guided by the founding principles that framed the European project from its start.

References

Bauer, D; & Papp, K (2009). "Book Review Perspectives: The Jevons Paradox and the Myth of Resource Efficiency Improvements". Sustainability: Science, Practice, & Policy.
Bowen H, (1953) Social responsibilities of the businessman, University of Iowa Press, 2013 Edition.
Chan S & Bencini J, Toolbox for multi-stakeholder climate partnerships German Development Institute, June 2018, p. 53.
Dayan S & Simonet G, (2020) Comité 21, S'adapter aux changements climatiques.
de Tocqueville A, On democracy in America (in 1835), 2002 University of Chicago Press.
Dugast C, (2019) Faire sa part? http://www.carbone4.com/publication-faire-sa-part/ Accessed 3 June 2020.
Gore, A. (1992) Earth in the Balance: Ecology and the Human Spirit. Houghton Mifflin, New York.
Heinberg R, There's no app for that (2017) Post Carbon Institute, Santa Rosa.
Heller A, (1976) The theory of needs in Marx, Allison & Busby, London.
Iselin R (2012) https://open.library.ubc.ca/cIRcle/collections/ubctheses/24/items/1.0073008 Accessed 3 June 2020.
Jackson T (2016 2nd Ed.) Prosperity without growth, Routledge, London.

Jevons, W (1865). The Coal Question; An Inquiry Concerning the Progress of the Nation, and the Probable Exhaustion of Our Coal Mines (1 ed.). London & Cambridge: Macmillan & Co.
Jonas H, (1984) The imperative of responsibility, University of Chicago Press.
Kusters C and Batjes K, (2017) Managing for Sustainable Development Impact. Practical Action Publishing, Rugby.
Lawn P & Clarke, M (2006) Measuring Genuine Progress: An Application of the Genuine Progress Indicator, Nova Science, Hauppauge.
Mackenzie M. K, Deliberation and Long-Term Decisions: Representing Future Generations. In The Oxford Handbook of Deliberative Democracy, Oxford University Press, 2018.
Maxwell, D., Owen, P., McAndrew. L, Muehmel, K. Neubauer, A. (2011), Addressing the Rebound Effect. European Commission DG Environment.
D Meadows et al. (1972) The limits to Growth, Club of Rome http://www.donellameadows.org/wp-content/userfiles/Limits-to-Growthdigital-scan-version.pdf Accessed 3 June 2020.
Mi, Z., Coffman, D. The sharing economy promotes sustainable societies. Nat Commun 10, 1214 (2019). https://doi.org/10.1038/s41467019-09260-4 Accessed 3 June 2020.
Monod, J-C, (2019) L'Art de ne pas être trop gouverné, Seuil, Paris.
OECD (2019), Measuring Distance to the SDG Targets 2019: An Assessment of Where OECD Countries Stand, OECD Publishing, Paris, https://doi.org/10.1787/a8caf3fa-en Accessed 3 June 2020.
Petersen A et al., Discrepancy in scientific authority and media visibility of climate change scientists and contrarians, Nature Communications, volume 10, 3502 (2019).
Piketty T, (2020) Capital and Ideology, Harvard University Press.
Randers J, (2012) 2052: A Global Forecast for the Next Forty Years, J Chelsea Green Publishing Co, Hartford.
Sale K, (1985) Dwellers in the Land: The Bioregional Vision, Sierra Club, Oakland.
Saez E & Sucman, G, (2019) The Triumph of Injustice: W. W. Norton New York.
Stiglitz J E, (2020) People, Power, and Profits: Progressive Capitalism for an Age of Discontent, WW Norton & Co, New-York.
Thompson, D. (2010) Representing future Generations: Political Presentism and Democratic Trusteeship. https://philpapers.org/rec/DENRFG Accessed 3 June 2020.
Thompson, Dennis. 2005. "Democracy in Time: Popular Sovereignty and Temporal Representation." Constellations 12 (2): 245–61.
Van Reybrouck D, (2018) Against elections, Seven Stories Press, New York.
Worms F, (2019) Pour un humanisme vital, Odile Jacob, Paris.

The manufacturer's authorised representative in the EU is Springer Nature Customer Service Centre GmbH, Europaplatz 3, 69115 Heidelberg, Germany. If you have any concerns regarding our products, please contact ProductSafety@springernature.com

Printed and bound by CPI Group (UK) Ltd, Croydon, CR0 4YY

30/09/2026

02225708-0009